Study Guidebook to accompany
BASIC STATISTICAL METHODS
Fourth Edition

Study Guidebook
to accompany

BASIC STATISTICAL METHODS

Fourth Edition

N. M. DOWNIE
Purdue University

HARPER & ROW, PUBLISHERS
New York Evanston San Francisco London

Sponsoring Editor: George A. Middendorf
Project Editor: Cynthia Hausdorff
Designer: T. R. Funderburk
Production Supervisor: Bernice Krawczyk

Study Guidebook to accompany Basic Statistical Methods, Fourth Edition

For information address Harper & Row, Publishers, Inc., 10 East 53rd Street, New York, N.Y. 10022.

ISBN 06-042734-5

CONTENTS

PREFACE

Each chapter in this study guide and workbook is associated with a corresponding chapter in Downie and Heath's *Basic Statistical Methods,* 4th ed. Before answering any questions in the workbook, study the material in the text carefully and thoroughly. Then, when you complete the items in the workbook you will recall and apply what you have learned, thus reinforcing your original learning.

Each chapter of the study guide has two basic parts. The first section of the chapter consists of objective questions in true-false, completion, short answer, or some other format. These questions cover in detail the material of the text. The second section of the chapter is made up of problems that will give you practice in applying statistics and solving problems with them. Using these study questions along with the text should make the beginning statistics course easier and, we hope, more enjoyable.

chapter 1
INTRODUCTION

I. TRUE-FALSE

1.	A major purpose of statistics is to make sense of a large collection of data.	1.	T	F
2.	Statistics is a phenomenon of the twentieth century.	2.	T	F
3.	The earliest type of statistics is descriptive.	3.	T	F
4.	A student's raw score on the College Boards is a statistic.	4.	T	F
5.	A score as in item 4 above is very meaningful.	5.	T	F
6.	An average indicates how the typical individual scored on a test.	6.	T	F
7.	A centile is a measure of how an individual performed in reference to others on a test.	7.	T	F
8.	A standard score is a measure of variability about the mean or average.	8.	T	F
9.	Studying two sets of measures taken on the same individuals against each other is correlational analysis.	9.	T	F
10.	Prediction is a major end of correlational analysis.	10.	T	F
11.	Determining the relationship between grades on the College Boards and freshmen grades is a typical example of correlational analysis.	11.	T	F
12.	One studies samples to make inferences about a population.	12.	T	F
13.	Inferential statistics came first in the historical development of statistics.	13.	T	F
14.	The use of small samples is very economical in terms of both time and money.	14.	T	F
15.	The major reason for not studying populations is that they are usually very large.	15.	T	F
16.	There are times when a group may be considered either a sample or a population.	16.	T	F
17.	Studying two groups to see if they differ significantly with respect to a certain trait is an example of inferential statistics.	17.	T	F
18.	Statistics is a basic tool of scientists.	18.	T	F
19.	Gamblers gave the greatest impetus to the development of statistics.	19.	T	F
20.	E. L. Thorndike was England's leading statistician.	20.	T	F
21.	Statistics entered the social sciences through the work of Gauss.	21.	T	F

22. Quetelet was the originator of correlational analysis. 22. T F
23. J. M. Cattell and his students were the first to use statistics widely in the United States. 23. T F
24. R. A. Fisher is known for his work with small sample statistics. 24. T F
25. The social sciences adopted small sample methodologies from agriculture. 25. T F
26. Galton developed the concept of the normal curve. 26. T F
27. Astronomers were involved in the early development of statistics. 27. T F

II. EXERCISES

1. List five uses of statistics in daily life that are already familiar to you.

   a.

   b.

   c.

   d.

   e.

2. From the above what can you infer about the day-to-day use of statistics?

3. Obtain three current issues of a research journal in your area of study and complete the following:

Issue	No. of Articles	No. of Articles Containing Statistics
1.		
2.		
3.		

4. Scrutinize the statistics in these articles. How do they differ from what you recorded in exercise 1 above?

5. Differentiate between inferential and descriptive statistics.

chapter 2

A REVIEW OF FUNDAMENTALS

I. COMPLETION

A negative number multiplied by a negative number results in a (1) ______ number, whereas the product of a negative and a positive number is always (2) ______.

When a number is multiplied by zero, the result is always (3) ______.

To subtract a negative number from a positive number, change the sign of the (4) ______ and (5) ______.

When 1 is divided by a number, the result is called the (6) ______ of the number. The easiest way to do this is to use (7) ______.

The square of a decimal is always (8) ______ than the decimal.

It is impossible to extract the square root of a (9) ______ number.

In the expression $5Y^3$, 5 is the (10) ______, Y stands for the (11) ______, and 3 is the (12) ______.

A proportion, p, is always part of a (13) ______; hence its maximum value is (14) ______.

Proportions are changed to percentages by multiplying each by (15) ______.

To assure adequate reliability, percentages should be based on at least (16) ______ cases.

Measures that increase by whole units are called (17) ______.

1. ______
2. ______
3. ______
4. ______
5. ______
6. ______
7. ______
8. ______
9. ______
10. ______
11. ______
12. ______
13. ______
14. ______
15. ______
16. ______
17. ______

Three examples of such measures are (18) ______, (19) ______, and (20) ______.

18. ______

19. ______

20. ______

In statistics we treat most measures as being (21) ______.

21. ______

The simplest type of measurement scale is called a (22) ______.

22. ______

After data have been ranked we have a (an) (23) ______ scale.

23. ______

A scale with equal units of measurement but without an absolute zero is called a (an) (24) ______ scale. An example of such a scale is (25) ______.

24. ______

25. ______

A meter stick is an example of a (26) ______ scale because it possesses (27) ______ and (28) ______.

26. ______

27. ______

28. ______

Suppose that we look at a new compact car. In saying that the car is a Mazda we use a (29) ______ scale. Adding that it is lighter than a Ford but heavier than another compact is an example of (30) ______ measurement. Testing its cooling system shows the temperature to be 165 degrees Fahrenheit. This is an example of a (an) (31) ______ scale. When it is said that gas consumption measured in miles per gallon is twice that of model B but only slightly better than compact C, we have first an example of a (an) (32) ______ and second a (an) (33) ______ scale.

29. ______

30. ______

31. ______

32. ______

33. ______

With an interval scale we can say that one object is (34) ______ taller than another; with a ratio scale an object may be (35) ______ taller.

34. ______

35. ______

Numbers describing samples are called (36) ______; those describing populations are called (37) ______.

36. ______

37. ______

Another name for population is (38) ______.

38. ______

The desirable statistical sample is referred to as a (39) ______ sample.

39. ______

The symbol Σ is always read as (40) ______.

40. ______

$X > Y$ means that X (41) ______ Y.

41. ______

$X \neq Y$ means that X (42) ______ Y.

42. ______

The symbol σ tells us that we are dealing with a (43) ______.

43. ____________

Parameter is to statistic as population is to (44) ______.

44. ____________

May the same group of individuals be considered either a sample or a population? (45) ______. An example of such a case is (46) ______.

45. ____________

46. ____________

II. EXERCISES

Complete the following operations:

1. 342 + (−13) + 15 + 280 − 13   1. ____________
2. 6.9 − (−8.3)   2. ____________
3. 8.2 − 12.6   3. ____________
4. (1.36) (−2.3)   4. ____________
5. 8.4327 ÷ (−2)   5. ____________
6. 1 ÷ 720   6. ____________
7. 1/2 + 1/7   7. ____________
8. 7/12 − 3/8   8. ____________
9. 7/12 × 5/8   9. ____________
10. 7/12 ÷ 5/8   10. ____________
11. $.000456^2$ 11. ____________
12. 1/888   12. ____________
13. $1/\sqrt{888}$ 13. ____________
14. $\sqrt{4858.35}$ 14. ____________
15. $\sqrt{4.85835}$ 15. ____________
16. $\sqrt{.000897}$ 16. ____________
17. $\sqrt{.0000897}$ 17. ____________
18. $\sqrt{1/109}$ 18. ____________

19. 4^4 19. ______

20. $4^4 - 6^3$ 20. ______

21. $(xy)(x^2y^3z)$ 21. ______

22. $(fx)(x)$ 22. ______

23. $\sqrt{4^4}$ 23. ______

24. 44 percent of 440 24. ______

25. .08 percent of 80 25. ______

26. 432 is what percent of 480? 26. ______

27. What proportion of 80 is 45? 27. ______

28. (23) (54) (90) (7) (0) 28. ______

Round to the nearest tenth:

29. 1.097 29. ______

30. .109 30. ______

31. 109.85 31. ______

32. 109.75 32. ______

33. .0898 33. ______

34. How many significant digits in exercise 29 above? 34. ______

35. How many significant digits in exercise 33 above? 35. ______

chapter 3
FREQUENCY DISTRIBUTIONS, GRAPHS, AND CENTILES

I. COMPLETION

Frequency distributions are constructed to make data more (1) ______.   1. ______

The high score in the distribution minus the lowest score plus 1 defines the (2) ______.   2. ______

In setting up a frequency distribution, the first step is to determine the (3) ______.   3. ______

If the highest score in a distribution is 77 and the lowest 7, the range is (4) ______.   4. ______

The typical frequency distribution has (5) ______ intervals.   5. ______

Dividing the range by (6) ______ is one method of determining the size of the class interval.   6. ______

Every interval has a (7) ______, (8) ______, and (9) ______.   7. ______

8. ______

9. ______

When building a frequency polygon, the ratio of the height of the curve to the width should be (10) ______.   10. ______

Values on the y axis are called (11) ______; another name for the x axis is (12) ______.   11. ______

12. ______

c.i. stands for (13) ______.   13. ______

When building a frequency polygon, the tally marks are always placed above the (14) ______ of the class intervals.   14. ______

If the frequencies of two or more distributions differ considerably and all distributions are to be plotted on the same axes, all sets of frequencies should first be changed to (15) ______.   15. ______

16. ______

The easiest way to change a frequency to a percentage is to take the (16) ______ of N, multiply this value by the (17) ______ of each interval, then multiply the result by (18) ______.   17. ______

18. ______

A distribution with an elongated tail extending to the left is said to have (19) ______ skew. 19. ______

A curve with cases piled up near the center is said to be (20) ______, whereas a curve in which the cases are widely distributed with a slight pile-up at the center is said to be (21) ______. 20. ______ 21. ______

The normal curve in reference to modes is (22) ______, has (23) ______ skew, is (24) ______, and is (25) ______ -kurtic. 22. ______ 23. ______ 24. ______ 25. ______

In terms of kurtosis, a distribution of scores that is homogeneous is described as (26) ______. 26. ______

In making histograms, we use the (27) ______ of the intervals. 27. ______

Graphs used to depict the percentage or proportion of the parts of a situation are called (28) ______ or (29) ______ graphs. 28. ______ 29. ______

In making an ogive curve, frequencies must first be changed to (30) ______. Then the tallies are placed above the (31) ______ of the intervals. 30. ______ 31. ______

C_{56} is defined as that (32) ______ in a distribution with (33) ______ of the cases below it. 32. ______ 33. ______

The midpoint of any distribution may be written as (34) ______, (35) ______, or (36) ______. 34. ______ 35. ______ 36. ______

Q_1 is equal to (37) ______; D_7 is equal to (38) ______. 37. ______ 38. ______

The shape of a distribution of centiles is (39) ______. 39. ______

Differences between centiles near the center of a distribution have (40) ______. 40. ______

When we say that 66 percent of the cases fall below a certain score, it follows that the score mentioned has a (41) ______ of 66. 41. ______

Another name for an ogive curve is (42) ______. 42. ______

From an ogive curve one may read both (43) ______ and (44) ______.

43. ______

44. ______

For each of the following give an actual illustration:

45. The normal curve.

46. A positively skewed distribution.

47. A negatively skewed distribution.

48. A bimodal distribution.

49. A platykurtic distribution.

50. A leptokurtic distribution.

Complete the exact limits and interval sizes for each of the following:

Interval	Exact Limits	Interval Size
51. 0–3		
52. 8–10		
53. 80–89		
54. (-5)–5		
55. .5–.75		

For each of the following give an appropriate interval size:

	High Score	Low Score	Interval Size
56.	20	8	
57.	40	2	
58.	128	8	
59.	+3	−3	
60.	69	19	

II. PROBLEMS

1. Below are the scores of a group of students on a statistics test:

89	72	67	61	52
86	71	66	61	50
84	70	65	59	50
84	70	65	59	48
82	70	65	58	46
78	69	65	57	43
76	68	64	57	42
76	68	64	56	41
75	68	64	54	40
72	67	63	53	36

   In the space at the right set up a frequency distribution for these data with the bottom interval as 35-39.

2. Using the same scores, set up another frequency distribution with the bottom interval as 36-40.

   a. Do the distributions differ? ________ Why?

b. What terms may be used to describe either of these distributions?

3. Using graph paper from the back of the book, construct a frequency polygon for each of the distributions made for problems 1 and 2 above.
4. Construct a histogram for the data in problem 1 above.
5. The following represent the scores of freshmen of two colleges on the SAT test of the College Boards:

	School A	School B
	f	*f*
760-779	4	2
740-759	18	10
720-739	26	7
700-719	42	8
680-699	63	12
660-679	56	14
640-659	52	2
620-639	30	4
600-619	20	3
580-599	12	2
560-579	6	1
540-559	5	2
520-539	0	2
500-519	0	0
480-499	2	0
460-479	0	1
	336	70

a. Plot these two distributions on the same axes.
b. Describe distribution A.

6. Treat the following as ungrouped data and find the median of each.

a. 77, 78, 80, 82, 83, 84, 86, 88
b. 62, 64, 65, 66, 67, 68, 70, 74, 79
c. 52, 56, 56, 56, 56, 58, 59, 62
d. 48, 49, 50, 52, 58, 59, 64, 65
e. 36, 38, 38, 38, 38, 38, 39, 39, 40

7. Below are the data from School A, problem 5:

	f
760-779	4
740-759	18
720-739	26
700-719	42
680-699	63
660-679	56
640-659	52
620-639	30
600-619	20
580-599	12
560-579	6
540-559	5
520-539	0
500-519	0
480-499	2
460-479	0
	336

For the above, set up the cumulative frequencies and cumulative proportions. Then on a piece of graph paper construct a cumulative proportion or ogive curve for the data.

8. From the curve constructed in problem 7 read the following centile points:

Median _____ Q_1 _____ C_{72} _____ D_8 _____

Q_3 _____ D_1 _____ C_{95} _____ D_2 _____

9. Compute the following centile points using the data in Problem 7 above. Compare your results with those obtained for problem 8.

Median _____ Q_1 _____ C_{72} _____ D_8 _____

Q_3 _____ D_1 _____ C_{95} _____ D_2 _____

10. Using the ogive curve constructed for problem 7 above read the centile rank for each of the following scores:

760 _____ 666 _____ 575 _____ 470 _____

11. Now compute the centile rank for each of the scores in problem 10.

760 _____ 666 _____ 575 _____ 470 _____

chapter 4
AVERAGES

I. COMPLETION

The arithmetic mean is obtained by (1) ______ the scores and dividing by the (2) ______.

1. ____________
2. ____________

The mean is that point in a distribution about which the sum of the (3) ______ equals (4) ______.

3. ____________
4. ____________

The word *moment* is synonymous with (5) ______.

5. ____________

$X - \overline{X}$ equals (6) ______. The sum of these after each has been squared is called (7) ______.

6. ____________
7. ____________

The sum of the squared deviations about the mean is (8) ______ than the sum taken about any other point in the distribution.

8. ____________

The average of four other means taken together is called the (9) ______ mean.

9. ____________

The sum of the scores in any distribution is equal to the (10) ______ multiplied by the (11) ______.

10. ____________
11. ____________

When a constant is subtracted from each score in a distribution, the mean (12) ______. On the other hand, if each score in a distribution is divided by the same constant, the mean is (13) ______.

12. ____________
13. ____________

The median is defined as that (14) ______ in a distribution with (15) ______ of the cases below it.

14. ____________
15. ____________

That score that occurs most frequently is called the (16) ______.

16. ____________

With grouped data, if the interval with the greatest frequency is the interval 70–79, the mode is (17) ______.

17. ____________

The mean is associated with a (18) ______ distribution, whereas the median is used with a (19) ______ distribution. 18. ______

19. ______

If the mean, median, and mode are identical, the distribution is (20) ______. 20. ______

A negatively skewed distribution will have the mean pulled toward the (21) ______ of the distribution. 21. ______

If the mean of a distribution is 28, the median 32, and the mode 36, we are dealing with a (22) ______ skewed distribution. 22. ______

As a statistic the mode is lacking in (23) ______, which means that (24) ______. 23. ______

24. ______

With ordinal data the appropriate average is the (25) ______; with interval data, the (26) ______. 25. ______

26. ______

The true mean of a distribution of test scores is 64.8. Actually a student obtained a mean of 64.6 by the grouping method for the same data. This difference between the two means is brought about by the (27) ______. In the long run the effect of such errors is (28) ______. 27. ______

28. ______

When we group data, we assume that the average of all scores in any given interval will be equal to the (29) ______ of the interval. Not meeting this assumption results in the (30) ______. 29. ______

30. ______

Each score in a distribution affects the size of the (31) ______. This makes it the most (32) ______ of the three measures of central tendency. 31. ______

32. ______

When one has a choice in selecting the measure of central tendency for use, whenever possible he should select the (33) ______. 33. ______

Items 34-42. Complete each of the following by use of the word *mean, median,* or *mode.*

To best represent the average salary of public school teachers in any given state (34) ______. 34. ______

To best represent the average salary of public school personnel (superintendents, principals, and teachers) in any given state (35) ______. 35. ______

On a test 4 out of 80 subjects do not finish in the allotted time. You want an average of the number of items completed (36) ______. 36. ______

The best figure to show the average income of all adult males in any given state (37) ______. 37. ______

654 high school students take a test that results in a bell-shaped distribution (38) ______. 38. ______

The heights of fifth-grade boys (39) ______. 39. ______

You have a frequency distribution in which the bottom five intervals are of \$500 each in size, the next five \$1,000 each, the next three \$10,000 each, and the last one over \$47,000 (40) ______. 40. ____________

To show what is the most popular type of car passing a certain point in a given period of time (41) ______. 41. ____________

To represent the "typical" individual, the so-called "average" man (42) ______. 42. ____________

II. PROBLEMS

1. The following scores were obtained by a group of students on a short quiz. Using formula 4.2 find the mean.

X	f
8	1
7	8
6	7
5	6
4	4
3	3
2	0
1	1
0	0

2. On a quiz the following scores were obtained by a group of students:

X	f
18	3
17	12
16	18
15	16
14	17
13	13
12	10
11	10
10	9
9	7
8	6
7	4
6	3
5	2
4	1
3	0
2	1

a. Find the mean using formula 4.3.
b. Find the median and mode for the same data.
c. Which average is most appropriate for these data?

3. On a final examination a statistics class produced the following distribution of scores:

	f
120-129	1
110-119	0
100-109	4
90- 99	9
80- 89	12
70- 79	8
60- 69	6
50- 59	4
40- 49	3
30- 39	2
20- 29	1

Compute the mean, median, and mode for the scores.

4. Given: 10, 12, 13, 14, 18, 20, 21, 22, 23, 25.
   a. Compute the mean.

   b. Add 20 to each score and recompute the mean.

   c. Divide each score by 2. Recompute the mean.

   d. Generalize from the above.

5. Given the following three means:
   $\bar{X} = 72, N_x = 50$
   $\bar{Y} = 80, N_y = 20$
   $\bar{Z} = 90, N_z = 10$
   The mean of all three groups, $\bar{X}_T$, taken together is:

6. Given two scores, 8 and 16.
   a. The arithmetic mean is

   b. The geometric mean is

   c. How is the geometric mean used? Can you find an example?

   d. The harmonic mean is

   e. How is the harmonic mean used? Can you find an example of this?

chapter 5
VARIABILITY

I. COMPLETION

As a measure of variability, the range is unsatisfactory because of its (1) ______. — 1. ______

In statistics x is a (2) ______ from the (3) ______. — 2. ______

3. ______

The sum of these deviations about the mean is always (4) ______ and the sum of each of these deviations squared is always (5) ______ than such sums taken about any other point. — 4. ______ 5. ______

When the deviations about the mean are summed disregarding signs and the sum is divided by N, the resulting statistic is the (6) ______. — 6. ______

To find the standard deviation, one needs the (7) ______. When this value is divided by N, the resulting statistic is called the (8) ______ or the (9) ______. — 7. ______ 8. ______

9. ______

When a standard deviation unit is taken on both sides of the mean, (10) ______ percent of the area is cut off; 2 standard deviations so taken include about (11) ______ percent of the area, and 3 standard deviations (12) ______ percent. — 10. ______ 11. ______ 12. ______

Standard deviation units are (13) ______ units of measurement. — 13. ______

When N is large, 400, about (14) ______ standard deviation units cover the range; when small, 30, (15) ______ standard deviation units cover the range. — 14. ______ 15. ______

If a constant of 8 is added to each score in a distribution, whose standard deviation is 12, the resulting standard deviation is (16) ______; however, if each score in this distribution is multiplied by 12, the standard deviation is (17) ______. — 16. ______ 17. ______

Sigma units and sigma are other names for (18) ______.

18. ______

s^2 is called the (19) ______.

19. ______

The mean-square is the same as the (20) ______.

20. ______

Q may be called either the (21) ______ or (22) ______.

21. ______

22. ______

If the variance is 99, s is (23) ______.

23. ______

$Q_3 - Q_1$ is the (24) ______.

24. ______

One Q taken on each side of the median cuts off (25) ______ percent of the cases.

25. ______

When Q and s are compared for the same data, (26) ______ will always be the larger.

26. ______

Standard deviation is used as a measure of variability when the (27) ______ is used as the measure of central tendency; (28) ______ is used when the median is used as the average. From this we conclude that s is used when a distribution is (29) ______, and Q when a distribution is (30) ______.

27. ______

28. ______

29. ______

30. ______

When a group has a large standard deviation for a given trait, the group is said to be (31) ______ in reference to the trait. On the other hand, a small standard deviation indicates (32) ______.

31. ______

32. ______

When the SS is 1280 and N is 24, the variance is (33) ______.

33. ______

If the variance for a set of data is 109 and N is 74, the SS for these data is (34) ______.

34. ______

The first moment about the mean is (35) ______. the second is the (36) ______, and the third is written as (37) ______.

35. ______

36. ______

37. ______

The formula for skew involves both the (38) ______ and the (39) ______ moments. When g_1 equals zero, skew is (40) ______, if greater than zero, skew is (41) ______, and if less than zero, the skew is (42) ______.

38. ______

39. ______

40. ______

41. ______

42. ______

The formula for kurtosis uses both the (43) ______ and (44) ______ moments. If g_2 is zero, the distribution is said to be (45) ______; when g_2 is greater than zero, the distribution is (46) ______; if negative, the distribution is (47) ______.

43. ____________

44. ____________

45. ____________

46. ____________

47. ____________

II. PROBLEMS

1. Below are the scores of 210 high school students on a clerical aptitude test:

	f
100-109	3
90-99	8
80-89	12
70-79	24
60-69	35
50-59	48
40-49	34
30-39	22
20-29	11
10-19	9
0-9	4
	210

Compute both the mean and standard deviation for these data.

2. Here are the scores of 60 students on a statistics test:

	f
75-79	1
70-74	0
65-69	3
60-64	4
55-59	8
50-54	12
45-49	9
40-44	9
35-39	7
30-34	4
25-29	2
20-24	1
	60

a. Calculate both the mean and the standard deviation for the above.

b. A standard deviation taken on each side of the mean cuts off approximately 68 percent of the cases. Since you now have both $\bar{X}$ and s for the two sets of data above, check this out.

3. Given:
   a. 10, 12, 14, 15, 18, 20, 25, 26, 28, 30
   b. 7, 8, 12, 14, 16, 18, 20, 21, 22, 22, 28, 30, 34
   c. 50, 54, 58, 60, 64, 68, 71, 75, 83, 88

For each of the above compute the following:

	$\overline{X}$	AD	s
a.			
b.			
c.			

4. Take the data in problem 3a above.
   a. Add a constant of 5 to each score. What is the standard deviation now? ______
   b. Multiply each score in the same distribution by 2. What is the standard deviation now? ______

5. Three groups of freshmen score as follows on an English placement test:

	N	$\overline{X}$	s
Group 1	80	154	10
Group 2	100	140	12
Group 3	60	160	8

   a. Compute the mean of the three groups combined. $\overline{X}_t$ = ________

b. Compute the standard deviation for the three groups combined. s_t = ________

6. Below are the same scores that appear in problem 2 above.

	f
75-79	1
70-74	0
65-69	3
60-64	4
55-59	8
50-54	12
45-49	9
40-44	9
35-39	7
30-34	4
25-29	2
20-24	1
	60

Calculate the quartile deviation for these data. Q = ________.

7. Here are the scores of 28 persons on a test.

48, 47, 46, 46, 45, 43, 42
41, 41, 40, 40, 40, 39, 39
35, 34, 33, 26, 25, 24, 23
22, 21, 20, 19, 17, 16, 14

Without grouping the data, find the quartile deviation. Q = ________

8. Given the following scores:

62	47
60	46
58	45
54	43
52	42
50	40
50	38
48	36
48	30
47	24

a. For the above calculate:

m_1

m_2

m_3

m_4

b. Using the data in part a above, find:

g_1

g_2

c. From the results obtained in 8b, how do you describe the skewness and kurtosis of this distribution?

Skewness

Kurtosis

chapter 6
STANDARD SCORES AND THE NORMAL CURVE

I. COMPLETION

A standard score is a raw score in (1) _____ form. 1. __________

When a person has a z score of 1.8, we know that he is 1.8 (2) _____ above the mean. 2. __________

z scores have a mean of (3) _____ and a standard deviation of (4) _____. 3. __________

4. __________

z scores are obtained by dividing the distance of a score from the (5) _____ by the (6) _____. 5. __________

6. __________

In a distribution that approximates a normal one, you would expect about half of the z scores to be (7) _____. 7. __________

In actual practice z scores tend to stay in the range (8) _____ 8. __________

When z scores are linearly transformed, each z score is multiplied by the (9) _____ and added to the (10) _____. 9. __________

10. __________

A z score of 1.6 is equal to a College Board score (CEEB) of (11) _____, a Naval score of (12) _____, and an ACT score of (13) _____. 11. __________

12. __________

13. __________

Standard scores make it possible to compare an individual's scores on (14) _____. 14. __________

If a group of scores is to be weighted, all data should be changed to (15) _____. 15. __________

A distribution of raw scores with a positive skew is transformed to z scores. The shape of this distribution will be (16) _____. 16. __________

There is (are) (17) _____ of normal curve(s). 17. __________

Since the tails of the normal curve never touch the base line, we say that the normal curve is (18) _____. 18. __________

The normal curve has (19) _____ skew, (20) _____ kurtosis, the maximum ordinate at the (21) _____, and (22) _____ mode(s).

19. __________

20. __________

21. __________

22. __________

Appendix B is based on the (23) _____ normal curve.

23. __________

If we cut off 30 percent of the area of the normal curve in one of its tails, we have also cut off 30 percent of the (24) _____.

24. __________

In a similar vein, if 20 percent of the area of a normal curve is below a certain score in a distribution, we can say that the (25) _____ of a person with this score is (26) _____.

25. __________

26. __________

One standard deviation taken on each side of the mean taken together cuts off (27) _____ percent of the cases, 2 standard deviations (28) _____ percent of the cases, and 3 standard deviations so taken (29) _____ percent of the cases. Above a z score of 2.8 we find (30) _____ percent of the area and (31) _____ percent of the area below it.

27. __________

28. __________

29. __________

30. __________

31. __________

When we normalize a distribution of scores, the new distribution has the (32) _____, (33) _____, and is based on the (34) _____.

32. __________

33. __________

34. __________

If a frequency polygon were drawn for each of the following, which ones do you think would take the shape of the normal curve?

Age at death of a sample of 1000 taken continuously in a city (35) _____.

35. __________

The circumference of the head (hat-band size) of a sample of 500 American males (36) _____.

36. __________

A distribution of the income of a sample of 600 20-year-old males in any state (37) _____.

37. __________

The weights of a sample of 200 freshly picked ripe tomatoes (38) _____.

38. __________

The weights of station wagons built by American auto manufacturers (39) _____.

39. __________

The scores of 400 university students on a personality or adjustment questionnaire (40) _____.

40. __________

II. PROBLEMS

1. Given a distribution with $\overline{X} = 70$ and $s = 8$. Change each of the following to *z* scores:

	z Score	Transformed Score
a. 70		
b. 78		
c. 62		
d. 90		
e. 50		
f. 75		

Next change each of the *z* scores to standard scores with a mean of 50 and a standard deviation of 10. Record your score in the column Transformed Score.

2. On a sixth-grade achievement test ten students score as follows:

	Scores			Standard Scores		
Student	Spelling	Language Usage	Reading	Spelling	Language Usage	Reading
A	118	53	120			
B	112	60	116			
C	82	39	77			
D	104	42	86			
E	126	54	106			
F	140	59	106			
G	143	62	112			
H	68	38	58			
I	78	33	81			
J	89	48	72			
$\overline{X}$*	100	45	80			
*s**	20	8	12			

*Based on a larger group.

a. Change all these raw scores to standard scores with $\overline{X} = 50$ and $s = 10$.
b. Average the standard scores for each student and then rank the students on overall performance, giving the best student a rank of 1.

Student	Average	Rank
A		
B		
C		
D		
E		
F		
G		
H		
I		
J		

3. It was decided that each student's grade in a certain course would be based 1/3 on a final exam, 1/6 on the first one-hour exam, 1/6 on the second, 1/6 on laboratory grades, and 1/6 on a cumulative score based on a series of short 10-item quizzes. The means and standard deviations are as follows:

	Final	Test 1	Test 2	Lab.	Quizzes
$\overline{X}$ =	120	54	76	88	100
SD =	15	8	10	4	8

A student receives these scores: final, 130; test 1, 56; test 2, 80; lab. 92; and quizzes, 122.

a. What is his weighted standard score for this course?

b. What letter grade would you give him for this course?

4. How much of the area of the normal curve lies between each of the following *z* scores and the mean?

		Area	Centile Equivalent
a.	.40		
b.	1.90		
c.	-.67		
d.	1.00		
e.	2.58		

What is the centile equivalent of each of these *z* scores?

5. In a normal distribution based on 800 cases, how many cases would you expect to find *above* each of the following z scores?

   a. -.05

   b. 1.645

   c. –.48

   d. 2.62

6. Another distribution based on 1500 cases has a mean of 76 and a standard deviation of 11. Assume a normal distribution.

   a. How many scores in this distribution would you expect to find above a raw score of 90?

   b. How many cases would you expect between raw scores of 70 and 90?

   c. How many cases would you expect to find below a raw score of 52?

7. The following distribution of scores was obtained on a mathematics test by a group of freshmen engineers.

	f
90-94	6
85-89	18
80-84	38
75-79	28
70-74	36
65-69	18
60-64	18
55-59	10
50-54	4
45-49	2
40-44	1
35-39	0
30-34	1

a. Find the mean and standard deviation of these scores.

$\overline{X}$ =

s =

b. By using the method illustrated in Table 6.2 in the text obtain the expected frequencies for these data.

c. Plot these frequencies along with the original ones on the same piece of graph paper.

chapter 7
CORRELATION—THE PEARSON r

I. COMPLETION

When we make a distribution of the age and weight of 100 individuals, we have a (1) ______ distribution.	1. ________
Correlation is a measure of (2) ______. It may or may not indicate (3) ______.	2. ________
	3. ________
A Pearson r of -.54 is (4) ______ one of .54.	4. ________
If one variable increases as the second variable decreases, we have an example of a (5) ______ relationship. When for any value of one variable, the second variable can take any value, there is (6) ______ relationship.	5. ________
	6. ________
The maximum value for a Pearson r is (7) ______; the minimum value (8) ______.	7. ________
	8. ________
When there is a perfectly positive relationship, the tallies on the scatterplot (9) ______; when there is no relationship, the tallies (10) ______.	9. ________
	10. ________
If the best and poorest students have the highest scores on a personality inventory and the average students the lower scores, the relationship is said to be (11) ______.	11. ________
A basic condition that must be met before a Pearson r is computed is that of (12) ______. By this is meant that the (13) ______ of the columns and rows fall along a (14) ______.	12. ________
	13. ________
When the tallies on a scatterplot go from the upper left to the lower right, the relationship is (15)______.	14. ________
	15. ________
When r is 1.00, there is (are) (16) ______ regression line(s).	16. ________
When the variances of all the columns on a scatterplot are approximately equal, the condition of (17) ______ is present.	17. ________

The basic Pearson r formula is written in terms of (18) _____ scores and is defined as the (19) _____.

18. _____

19. _____

Dividing the sum of the cross-products by N results in the (20) _____.

20. _____

The variance of the difference between two measures is equal to the (21) _____ of the first, plus the (22) _____ of the second, minus the (23) _____ term.

21. _____

22. _____

23. _____

When a research worker computes a Pearson r, it is essential that he always make a (24) _____ to ascertain the presence or absence of (25) _____.

24. _____

25. _____

The greater the range, the larger the (26) _____.

26. _____

The correlation between age and height when determined on a sample of sixth graders would be (27) _____ than a similar coefficient computed on children in grades 4 through 7.

27. _____

The size of a Pearson r is more dependent on the (28) _____ than on the (29) _____ of cases.

28. _____

29. _____

If a Pearson r is computed for data that are not essentially linear, the r so computed will be a (an) (30) _____ of the true relationship.

30. _____

When curvilinearity is present the correct correlation coefficient to be used is the (31) _____.

31. _____

Often there is a correlation between two variables because both variables are (32) _____.

32. _____

Other times a high correlation coefficient appears between two unrelated variables because of the effects of (33) _____.

33. _____

Very simply, reliability means (34) _____. If an achievement or an intelligence is well made, one would expect to find reliability coefficients above (35) _____.

34. _____

35. _____

When we correlate scores on a test of ability with the ratings of supervisors on a job, we are studying the (36) _____ of the test. The ratings are called the (37) _____. Such correlations are (38) _____ than reliability coefficients, typically being (39) _____.

36. _____

37. _____

38. _____

39. _____

When we are talking about the statistical significance of a Pearson r, we are describing whether or not it differs from (40) ______; or in other words, whether there is a (41) ______ relationship between the two variables. | 40. ______ | 41. ______

Suppose that in a correlation problem both the X and Y values are coded by subtraction of a constant from each score in each. The Pearson r for the coded data will (42) ______ that for the uncoded data. | 42. ______

If data do not possess homoscedasticity, one encounters difficulty when using such data for (43) ______. | 43. ______

For each of the following situations made up of pairs of variables, describe in several words the type of relationship that you think exists, such as low positive, high negative, curvilinear, etc.:

Size of cities and the number of elementary schools (44) ______. | 44. ______

Mental ability scores and first semester grades for a group of college freshmen (45) ______. | 45. ______

Mental ability scores and the grades of graduating seniors in college (46) ______. | 46. ______

The ability to run the 100-yard dash and the ability to add 2-digit numbers (47) ______. | 47. ______

Social status and the proportion of income spent on food for a group of families (48) ______. | 48. ______

In general, the supply of a product and its price (49) ______. | 49. ______

Ages of individuals and scores on tests of physical strength (50) ______. | 50. ______

Height and age of adult males (51) ______. | 51. ______

Height and age of boys age 12 (52) ______. | 52. ______

Height and age of boys ages 8 to 13 (53) ______. | 53. ______

Grades in English courses and grades in shop courses such as woodworking (54) ______. | 54. ______

Grades in Spanish and grades in French (55) ______. | 55. ______

Grades in English and grades in music (56) ______. | 56. ______

Horsepower of automobiles and miles per gallon obtained (57) ______. | 57. ______

II. PROBLEMS

1. The following scores were obtained by 15 individuals on an intelligence test (X) and on a test of spatial relations (Y).

Ind.	X	Y
1	64	48
2	36	20
3	42	36
4	41	38
5	28	19
6	56	42
7	52	46
8	38	33
9	59	41
10	39	21
11	49	36
12	54	50
13	29	18
14	38	28
15	44	38

a. Calculate the Pearson r for the above data using formula 7.5.

b. Find the mean and standard deviation for each distribution.

$\overline{X}$

$\overline{Y}$

s_x

s_y

2.

Ind.	Initial Score	Final Score
1	10	12
2	8	7
3	14	13
4	6	8
5	4	7
6	8	6
7	7	6
8	3	4
9	7	9
10	10	11

Using formula 7.5 find the correlation between the two sets of scores.

3.

Ind.	Initial Score	Final Score
1	10	12
2	8	7
3	14	13
4	6	8
5	4	7
6	8	6
7	7	6
8	3	4
9	7	9
10	10	11

These are the same scores that appear in problem 2. Find the Pearson r by the method of differences, formula 7.7.

4.

Ind.	Initial Score	Final Score
1	10	12
2	8	7
3	14	13
4	6	8
5	4	7
6	8	6
7	7	6
8	3	4
9	7	9
10	10	11

Again these are the same scores as in problem 2. Find the Pearson r by the method of sums, formula 7.9.

5. Below are the scores of 60 individuals on two forms of a short test of mental ability:

Form AS	Form BS	Form AS	Form BS	Form AS	Form BS
15	17	8	7	19	18
16	15	14	14	11	14
17	10	10	13	16	18
16	16	15	19	15	18
16	17	4	7	20	22
19	21	22	20	12	13
15	19	12	6	16	18
13	15	21	20	5	7
17	18	12	14	18	21
17	17	8	7	15	11
19	20	12	13	17	23
14	12	15	18	11	14
14	16	18	19	13	16
10	13	12	14	11	8
12	8	13	16	7	10
15	13	16	18	7	12
17	20	18	23	21	24
11	9	7	12	18	17
16	16	10	9	8	9
10	10	14	16	20	18

a. Using one of the scatterplots in the back of the book set up a bivariate distribution for these data. By examining the scatterplot, what inference can you make about the relationship between the two variables?

b. By any method calculate the Pearson r between the two sets of scores.

chapter 8
OTHER CORRELATIONAL TECHNIQUES

I. COMPLETION

Responses to a test item such as "right-wrong" constitute a (1) _____. 1. _____

When one uses the point-biserial r, one of the variables is (2) _____ and the other a (3) _____. 2. _____ 3. _____

Essentially the r_{pbis} is a (4) _____. 4. _____

In the formula for the point-biserial r, p stands for (5) _____ and q is (6) _____. 5. _____ 6. _____

Graphs used for rapidly estimating statistics are called (7) _____. 7. _____

The biserial r differs from the point-biserial r in that one of the variables in the former is a (an) (8) _____. 8. _____

In the formula for the biserial correlation coefficient, y stands for (9) _____. 9. _____

When both biserial and point-biserial coefficients are computed for the same data, the (10) _____ is the larger. The (11) _____ is the more reliable of the two. 10. _____ 11. _____

This difference between a forced dichotomy and a real dichotomy is often (12) _____. 12. _____

Another name for the fourfold coefficient is the (13) _____ coefficient. When this coefficient is used, both variables are (14) _____. 13. _____ 14. _____

Phi reaches its maximum size when (15) _____. 15. _____

When both of two variables have been forced into dichotomies, the (16) _____ coefficient may be used. 16. _____

The tetrachoric correlation coefficient is obtained today mostly from a (17) _____ which is entered with the value (18) _____. 17. _____ 18. _____

Compared to the Pearson r, the tetrachoric r is much less (19) ______.

19. ______

Two coefficients that are much less useful today than they were before the days of the computer are (20) ______ and (21) ______.

20. ______

21. ______

When data are nonlinear and when a Pearson r is computed for such data, the obtained r is a (an) (22) ______ of the real relationship. The coefficient that should be used is the (23) ______ or (24) ______.

22. ______

23. ______

24. ______

When we are solving for eta we must get the so-called "between" (25) ______ and the (26) ______ sum-of-squares.

25. ______

26. ______

Eta and r are equal when (27) ______; eta is larger than r when (28) ______. The difference between eta and r then gives an indication of departure from (29) ______.

27. ______

28. ______

29. ______

The sign of eta is (30) ______.

30. ______

A partial r removes the effect of a (31) ______ variable from the correlation between (32) ______.

31. ______

32. ______

A multiple correlation coefficient indicates the correlation between one variable and the (33) ______ of two or more variables.

33. ______

The Spearman rho may be used with any data that can be reduced to (34) ______.

34. ______

For all practical purposes a rho of .89 is the equivalent of an r of (35) ______.

35. ______

Rho is most practical when sample sizes are (36) ______.

36. ______

A coefficient that is the equivalent of the average of a group of Spearman rhos is the (37) ______.

37. ______

Below is a group of situations that require a correlation coefficient. For each one, select what you feel is the most appropriate coefficient.

Responses to a test item and the total scores on a test (38) ______.

38. ______

Pulse rate and the intake of varying amounts of drugs for a group of small vertebrates (39) ______.

39. ______

Sex of respondents and the tendency to agree or disagree with an item on an attitude scale (40) ______. 40. ______

The reliability of a group of judges in a bathing beauty contest (41) ______. 41. ______

Grades in physics and grades in mathematics with the effects of mental ability removed (42) ______. 42. ______

First semester college grades and three predictors, the verbal and mathematics scores on the SAT of the College Boards and high school rank (43) ______. 43. ______

Hand strength and chronological age (44) ______. 44. ______

Composite scores on an aptitude test and completion of a course of training (45) ______. 45. ______

Scores on the SAT Verbal of the College Boards and first semester grades for 645 university freshmen (46) ______. 46. ______

Placement in the top or bottom half of a distribution of test scores and responding correctly to an item scored dichotomously (47) ______. 47. ______

Scores on a personality test and scores on a test of mental ability (48) ______. 48. ______

II. PROBLEMS

1. In the first column of the data below are the scores of 100 individuals on Test Y. The next two columns show the number who answered another test item right or wrong. For example, of those who scored 20 on Test Y, 3 persons answered the item correctly and no one missed it. Calculate the point-biserial r between the scores on Test Y and the responses to the other item. Use formula (8.1).

Y	f_r	f_w
20	3	0
19	4	0
18	6	2
17	8	1
16	7	3
15	6	4
14	4	5
13	3	6
12	2	5
11	1	8
10	0	9
9	1	12

2. The following data are similar to those in problem 1 above.

Scores	Right	Wrong
100-109	12	1
90-99	18	3
80-89	16	7
70-79	16	9
60-69	10	10
50-59	8	12
40-49	6	10
30-39	2	12
20-29	0	14
10-19	2	12

Compute the r_{pbis} for these data using formula 8.2.

3. Compute r_{bis} for the data in problem 2.

4. Given the following responses to a test item, compute the phi coefficient:

	Right	Wrong
High Group	55	45
Low Group	35	65

5. In responding to an item on an attitude scale, 80 out of 200 males and 50 out of 100 females agreed with the item. By the use of the phi coefficient determine if the sex of the individual and type of response are related.

6. Using Appendix H in your text obtain the tetrachoric correlation coefficient for the data in problems 4 and 5 above.

7. In a given situation the correlation between intelligence test scores and scores on a language test was found to be .70. The correlation of these intelligence test scores and scores on an arithmetic test was found to be .60. The correlation between these language and arithmetic scores was .45. What is the correlation between the last two variables with the effects of intelligence partialled out?

8. Using the data in problem 7, find the multiple correlation coefficient between intelligence test scores and the combined effects of language and arithmetic test scores.

9. Below are the scores of 60 individuals on two forms of the same test:

Form AS	Form BS	Form AS	Form BS	Form AS	Form BS
15	17	8	7	19	18
16	15	14	14	11	14
17	10	10	13	16	18
16	16	15	19	15	18
16	17	4	7	20	22
19	21	22	20	12	13
15	19	12	6	16	18
13	15	21	20	5	7
17	18	12	14	18	21
17	17	8	7	15	11
19	20	12	13	17	23
14	12	15	18	11	14
14	16	18	19	13	16
10	13	12	14	11	8
12	8	13	16	7	10
15	13	16	18	7	12
17	20	18	23	21	24
11	9	7	12	18	17
16	16	10	9	8	9
10	10	14	16	20	18

For the above data, construct a scatterplot, putting the AS scores on the Y axis and the BS scores on the X axis. Then compute the eta coefficient between Y and X.

These same data appeared as problem 5 in Chapter 7. If you solved that problem for the Pearson r, compare your eta coefficient with that r.

10. Below are the scores of 15 individuals on two tests:

X	Y
80	68
24	22
36	18
62	58
48	39
72	68
38	46
55	62
72	60
64	68
63	62
59	49
28	18
36	34
70	59

Calculate the Spearman rank-order correlation for these data.

11. Using the data in problem 10 above, compute a tau coefficient.

12. Four judges rank 8 projects as follows:

Project	Judge 1	2	3	4
1	7	8	8	7
2	8	6	7	8
3	5	7	6	5
4	1	2	1	2
5	6	5	5	6
6	4	3	4	4
7	2	1	3	1
8	3	4	2	3

Compute W for the above data.

chapter 9
LINEAR REGRESSION

I. COMPLETION

Before a Pearson r is computed one must have a (1) _____ relationship between two variables.

The predictor is known as the (2) _____ variable and that which is predicted the (3) _____ variable.

In the equation for a straight line the a coefficient is the (4) _____ and the b coefficient the (5) _____ of the line.

The b coefficient is the ratio of the change in (6) _____ in relation to the change in (7) _____.

The slope of a line may be either (8) _____ or (9) _____. If a line goes from the upper left quadrant of a scatterplot to the lower right, the slope of this line is (10) _____.

The difference between the obtained and predicted score is the (11) _____.

A regression line is that line about which the (12) _____ of the errors of prediction is at a (13) _____. Such a line is called a (14) _____.

There are (15) _____ coefficients. One is the coefficient related to predicting (16) _____ from (17) _____ and the other in predicting (18) _____ from (19) _____.

1. _____________
2. _____________
3. _____________
4. _____________
5. _____________
6. _____________
7. _____________
8. _____________
9. _____________
10. _____________
11. _____________
12. _____________
13. _____________
14. _____________
15. _____________
16. _____________
17. _____________
18. _____________
19. _____________

The b coefficient is the ratio of the (20) ______ to the (21) ______.

20. ____________

21. ____________

The product of the two b coefficients equals (22) ______.

22. ____________

The two regression lines cross at a point equal to (23) ______ and (24) ______.

23. ____________

24. ____________

If b_{yx} equals 1.20 and b_{xy} equals .30, then r is (25) ______.

25. ____________

A second formula for the b coefficient is the ratio of the two (26) ______ multiplied by (27) ______.

26. ____________

27. ____________

Items 28-32 pertain to the data below.

Given: $\bar{X} = 40$ $\bar{Y} = 70$

$s_x = 10$ $s_y = 5$

If $r = .00$, what is Y' for an X of 30? (28) ______.

28. ____________

If $r = .70$, what is Y' for an X of 30? (29) ______.

29. ____________

If $r = +1.00$, what is Y' for an X of 30? (30) ______.

30. ____________

If $r = -1.00$, what is Y' for an X of 30? (31) ______.

31. ____________

If $r = .40$, what is X' for a Y of 80? (32) ______.

32. ____________

Unless $r = +1.00$, we always have a (an) (33) ______ when we predict.

33. ____________

When $r = .00$, the predicted values of Y fall on a line parallel to the (34) ______ cutting the Y axis at (35) ______. Or stated in other words, when $r = .00$, (36) ______ equals (37) ______.

34. ____________

35. ____________

36. ____________

37. ____________

The maximum size of the standard error of estimate is (38) ______ and the minimum (39) ______.

38. ____________

39. ____________

The standard error of estimate is a measure of the (40) ______ about the (41) ______.

40. ____________

41. ____________

A regression line is actually a line of (42) ______.

42. ____________

Since the standard error of estimate is actually a (43) ______, we can interpret it by saying that for any given X value the chances are 2 out of 3 that an individual's Y score will fall in the band (44) ______.

43. ____________

44. ____________

The (45) ______ the standard error of estimate, the better we can predict.

45. ____________

In order to interpret the standard error of estimate, which other condition basic to a Pearson r has to be present? (46) ______.

46. ____________

When $r = 1.00$, there is (47) ______ regression.

47. ____________

If Mr. D scored 2½ standard deviations above the mean on test A and if he later took test B, an equivalent test, we would expect him to fall (48) ______ to the mean of B than he did to the mean of A.

48. ____________

Another name for the predicted variable is the (49) ______.

49. ____________

The standard error of estimate tends to be (50) ______ throughout the range.

50. ____________

Prediction may be enhanced by the use of (51) ______ predictors.

51. ____________

When we write $R_{1 \cdot 2345}$, the number 1 in the subscript stands for the (52) ______ and the other elements in the subscript are the (53) ______. The efficiency of prediction may be increased by the addition of (54) ______ variables.

52. ____________

53. ____________

54. ____________

II. PROBLEMS

1. Solve each of the following for Y':

	X	a	b	Y'
a.	6	3	4	
b.	−8	4	2	
c.	12	8	½	
d.	9	2	−1½	

2. The following scores were obtained by 15 individuals on an intelligence test (X) and a test of spatial relations (Y). (The same data appeared in problem 1, Chapter 7.)

X	Y	X	Y
6	4	5	4
3	2	3	2
4	3	4	3
4	3	5	5
2	1	2	1
5	4	3	2
5	4	4	3
3	3		

a. Compute a Pearson r for these data.

b. Set up the regression equation for predicting Y from X.

c. Set up the regression equation for predicting X from Y.

d. Compute the standard error of estimate when predicting Y from X.

e. Make a statistical test of the correctness of the two regression equations.

f. Plot these two regression lines on a piece of graph paper.

g. How does this indicate the correctness of your work?

h. There are two regression lines, but in actual practice a personnel worker is only concerned with one of them. Why?

3. Given the following data:
$\overline{X} = 500$ $\quad s_x = 100$ $\quad r_{xy} = .50$
$\overline{Y} = 4.0$ $\quad s_y = .70$

a. Using formula 9.9 in your text obtain Y' for an X of 600.

b. Compute the standard error of estimate for predicting Y from X.

4. In the following Y represents sales index and X represents scores on a sales attitude test.

X	Y
40	12
38	8
62	22
54	16
60	17
30	8
55	19
58	18
53	12
46	14

a. Set up the regression equation for predicting Y from X.

b. Compute the standard error of estimate associated with predicting Y from X.

c. When X is 48, what is Y'?

d. Interpret this using the standard error of estimate obtained in b above.

chapter 10

PROBABILITY AND THE BINOMIAL DISTRIBUTION

I. COMPLETION

The highest probability value indicates (1) ______. 1. ______

If the probability of a particular horse winning a race is 10 to 1, we would say that he is a (2) ______ shot. 2. ______

If the probability of the occurrence of an event is 7/8, the chance that this event will *not* occur is (3) ______. 3. ______

The probability of the occurrence of three separate events is the (4) ______ of their (5) ______. 4. ______

5. ______

In probability each event is described as being (6) ______. 6. ______

The probability of drawing three consecutive sevens, with replacement, from a well-shuffled deck of cards is (7) ______. 7. ______

As M increases and p equals 5, the shape of the binomial curve approaches that of the (8) ______ but never reaches it because the binomial curve is (9) ______. 8. ______

9. ______

The result obtained by tossing two coins is similar to that obtained by squaring the (10) ______. 10. ______

Pascal's triangle is properly used when p = (11) ______. 11. ______

If 12 coins are randomly tossed, the sum of all possible probabilities is (12) ______. 12. ______

(13) ______ denotes the (14) ______ of the binomial and Npq denotes (15) ______. 13. ______

14. ______

15. ______

Suppose that a student took a 200-item T-F test and knew absolutely nothing about it; his chance score would be (16) ______. 16. ______

Given $p = .6$, $q = .4$, $N = 100$, the variance of the binomial is (17) ______. 17. ______

In using the binomial to solve a problem, a student uses the term $z = (19 - 12)/3$ where 19 is a score, 12 the mean, and 3 the standard deviation. What should he have used? (18) ______. 18. ______

II. SHORT ANSWER

1. In a box there are 10 balls, 6 reds, 3 whites, and 1 blue.

   a. What is the probability of drawing a white ball? a. ______

   b. A white or a red ball? b. ______

   c. A white, red, or blue ball? c. ______

   d. A red ball followed by a blue ball, with replacement? d. ______

   e. Three white balls in a row, with replacement? e. ______

   f. What is the probability of the fifth ball being red when the first four draws are red? Assume replacement. f. ______

   g. What is the probability of drawing a red, blue, and white ball in that order, with replacement? g. ______

2. What is the probability of the occurrence of each of the following events?

   a. Drawing the ace of spades from a well-shuffled deck of cards? a. ______

   b. Drawing any ace? b. ______

   c. Drawing a club? c. ______

   d. Drawing the seven, eight, or nine of clubs? d. ______

   e. Drawing a spade on the fourth draw when the first two draws were hearts and the third a club. Each card was replaced after being drawn. e. ______

   f. Drawing the card named in part e above with no replacement? f. ______

   g. Obtaining 5 consecutive heads when tossing coins? g. ______

   h. Ten coins are tossed. What is the probability that they will be either all heads *or* all tails? h. ______

   i. When 10 coins are tossed what is the probability of getting 8 or more heads? i. ______

   j. In part i above, what is the probability of obtaining exactly 8 tails? j. ______

   k. What is the probability of getting an odd number when a die is tossed? k. ______

l. What is the probability of throwing a pair of six-spots with fair dice? l. ______________

m. What is the probability of obtaining a score of 20 on a 40-item true-false test when you know absolutely nothing about the subject? m. ______________

III. PROBLEMS

1. Take 8 similar coins and toss them 128 times. Record your results below:

8 Heads	0 Tails	________
7 Heads	1 Tail	________
6 Heads	2 Tails	________
5 Heads	3 Tails	________
4 Heads	4 Tails	________
3 Heads	5 Tails	________
2 Heads	6 Tails	________
1 Head	7 Tails	________
0 Heads	8 Tails	________
		________
		$\Sigma = 128$

Compare your results with what might be expected as read from Pascal's triangle.

2. Six dice are tossed.
   a. What is the probability of obtaining 4 six-spots?

b. What is the probability of obtaining *4 or more* six-spots?

3. Twelve coins are tossed 4096 times.
   a. Use the binomial expansion to find the probability of obtaining exactly 10 heads.

   b. Use the binomial expansion to find the probability of obtaining 10 or more heads.

   c. Since Np is greater than 5, solve the above using the formulas for the mean and standard deviation of the binomial, again finding the probability of obtaining exactly 10 heads.

   d. As in c above, find the probability of obtaining 10 or more heads.

4. Suppose that you have an 80-item multiple-choice test in statistics, each item being made up of four alternatives.
   a. What is the so-called chance score on this test?

b. Suppose that pure chance is operating. What is the probability of a person's getting a score of 30 or higher on this test?

c. Of getting a score of exactly 30?

5. A study was made using four types of perfume. One was a cheap brand and the others more expensive. Each subject was to smell all four and then select the cheap brand. Of 120 subjects who participated in the study, 37 correctly identified the cheap brand. Do these results differ from what you would expect by chance?

chapter 11
SAMPLING

I. COMPLETION

When we make a statement about a population from data obtained on a sample, we are making a statistical (1) ______. — 1. ____________

There are two basic types of sample, (2) ______ and (3) ______. — 2. ____________ 3. ____________

The use of students in an elementary psychology class as subjects in an experiment is an example of a (an) (4) ______ sample. — 4. ____________

When the sample contains the same percentages in the subgroups as in the population, this is called a (5) ______ sample. — 5. ____________

Nonprobability samples are used chiefly because they are (6) ______ when compared to probability samples. — 6. ____________

Statistically the ideal sample is the (7) ______ sample. In such a sample every individual has a (an) (8) ______ chance of being part of the sample. Unless a sample is drawn like this, it is most likely to be a (an) (9) ______ sample. — 7. ____________ 8. ____________ 9. ____________

Taking a random sample of individuals who have registered automobiles will result in a (an) (10) ______ of the residents of a state. — 10. ____________

When the individuals who make up the subclasses of a sample are drawn randomly we have a (an) (11) ______ sample. — 11. ____________

A more practical type of random sampling when the population is very large is the (12) ______ sample. — 12. ____________

Universe is another name for (13) ______. — 13. ____________

In statistics, the population is (14) ______ by the research worker. — 14. ____________

A useful device in setting up a random sample is a (an) (15) ______. — 15. ____________

A distribution of sample means about the population mean is called a (16) _____ distribution. Variability of sample means about this parameter mean is expressed by the (17) _____, which is defined as the (18) _____ divided by the (19) _____.

16. __________

17. __________

18. __________

19. __________

The distribution of samples about the population mean takes the shape of the (20) _____ distribution.

If we have a parameter with a mean of 70 and a standard error of 3, we can say that the chances are (21) _____ that another sample of the same size drawn from the same population will fall within the band (22) _____.

Standard errors are essentially measures of the (23) _____ of a statistic.

The size of the standard error of the mean is (24) _____ proportional to $\sqrt{N}$ and (25) _____ proportional to the standard deviation.

20. __________

21. __________

22. __________

23. __________

24. __________

25. __________

A sample mean is a (an) (26) _____ estimate of the parameter mean, whereas the sample standard deviation is a (an) (27) _____ estimate of the population standard deviation.

26. __________

27. __________

$N - 1$ is used in the denominator of the formula for the sample standard deviation to (28) _____.

Every statistic has a (an) (29) _____.

28. __________

29. __________

The standard error of the median is approximately (30) _____ larger than the standard error of the mean. This means that the mean is a more (31) _____ statistic.

30. __________

31. __________

In statistics q equals (32) _____, or $p + q$ equals (33) _____.

32. __________

33. __________

Multiplying the (34) _____ by 2.58 (for a large sample), adding this product to the mean, and then subtracting it from the mean will determine the limits of the (35) _____ interval. We can now be (36) _____ confident that the (37) _____ falls within this band.

34. __________

35. __________

36. __________

37. __________

The 99 percent confidence interval is (38) _____ than the 95 percent interval.

We study samples in order to make (39) _____ about the (40) _____.

38. __________

39. __________

40. __________

If the sample used in a study is a (41) _____ sample, we can make (42) _____ valid inferences about the population.

41. ____________

42. ____________

II. PROBLEMS

1. By using the table of random numbers in the Appendix of the textbook, pull ten samples of five cases each from scores below. Record your results in the space below.

15	12	15	12
16	15	16	17
17	18	11	5
16	16	16	16
16	13	10	10
19	16	8	7
15	18	14	14
13	8	10	13
17	10	15	9
17	14	4	7
19	19	22	7
14	11	12	21
14	16	21	18
10	15	12	14
12	20	8	7

	Sample									
	1	2	3	4	5	6	7	8	9	10
1.										
2.										
3.										
4.										
5.										
$\overline{X}$										

a. Compute the mean of each sample and record the result above.

b. Next compute the mean of the ten sample means. What is this the best estimate of?

c. Find the standard deviation of these ten sample means about the mean of the sample means.

d. What name is given to this statistic?

e. Suppose that you had pulled samples of size 10 instead of 5. How would this have affected the mean of the sample means? The standard error of the mean?

2. Outline how you would draw a random sample of 80 fifth-grade boys from a school system that has in it 12 elementary schools, each with a single fifth grade.

3. Given the following data.

$\bar{X}_1 = 86$ $\bar{X}_2 = 82$

$s_1 = 11$ $s_2 = 8$

$N_1 = 145$ $N_2 = 82$

a. For each, compute the standard error of the mean.

$s_{\bar{x}_1}$

$s_{\bar{x}_2}$

b. Set up the 95 percent and 99 percent confidence interval for each of the following:

s_{x_1}

95 percent confidence interval

99 percent confidence interval

s_{x_2}

95 percent confidence interval

99 percent confidence interval

c. In your own words describe what the 95 percent confidence interval means. Use the data for part a above.

4. For each set of data in problem 3, compute the standard error of the median.

s_{mdn_1} =

s_{mdn_2} =

Compare your results with the standard error of the mean for each group. What does this tell you about these two statistics?

5. Suppose a normal distribution with mu = 80 and sigma = 20.
   a. What is the probability of drawing a sample of size 100 from this population with a mean of 82 or higher?

   b. What is the probability of drawing a sample of the same size with a mean of 75 or less?

chapter 12

TESTING HYPOTHESES: TESTS RELATED TO MEANS

I. COMPLETION

A sample has a mean of 103, which is a (1) _____. 1. ____________

Every statistical test starts with a (an) (2) _____. This is accompanied by a (an) (3) _____. 2. ____________ 3. ____________

If we reject H_0, we accept the (4) _____. 4. ____________

A research worker decides to use the 1 percent level in making tests of significance. This is called the (5) _____. 5. ____________

If H_0 is not rejected when it is true, we say that a Type (6) _____ error has been made. If rejected when actually false, a Type (7) _____ error has been made. 6. ____________ 7. ____________

Increasing the size of the sample tends to reduce Type (8) _____ error. 8. ____________

When z is used in testing an hypothesis and a one-tailed test is made, the critical value at the 5 percent level is (9) _____ and the 1 percent value (10) _____. With a two-tailed test the corresponding values are (11) _____ and (12) _____, respectively. 9. ____________ 10. ____________ 11. ____________ 12. ____________

If H_0 is rejected at the 1 percent level, the chances are (13) _____ that the results are significant. 13. ____________

When a one-tailed test is made, H_0 has (14) _____ alternate hypotheses. 14. ____________

Values of t and z are the same when (15) _____. 15. ____________

A distribution with 7 scores has a *df* of (16) _____. 16. ____________

In a problem with 47 pairs of correlated data, there are (17) _____ degrees of freedom. 17. ____________

Sample variances are pooled to get a (an) (18) _____ estimate of the (19) _____ variance.

18. __________

19. __________

t is defined as the ratio of (20) _____ to the (21) _____.

20. __________

21. __________

When the variances of two samples do not differ statistically, we say that we have (22) _____ of variance. The test used for this is the (23) _____, defined as the ratio of (24) _____ to (25) _____.

22. __________

23. __________

24. __________

25. __________

A second condition that has to be met when the F test is used is (26) _____.

26. __________

Given $s_1^2 = 120.4, N_1 = 18, s_1^2 = 84.8, N_2 = 16$. F is (27) _____.

27. __________

With an F of this size H_0 related to the two variances would be (28) _____.

28. __________

When the assumptions related to a statistical test are violated and the test is relatively unaffected, we say that the test is (29) _____.

29. __________

II. PROBLEMS

1. On a test with an established mean IQ of 100 and a standard deviation of 15, a sample of 16 has a mean IQ of 107. Does the mean of this group differ significantly from that of the norm group?

2. A sample of 36 university freshmen has a mean of 525 on the SAT verbal. Does this mean differ significantly from that of the norm group that has a mean of 500 and a standard deviation of 100?

3. The following results were obtained by the performance of males and females on a design preference test.

	N	$\overline{X}$	s
Males	145	64.6	8.3
Females	90	68.9	9.1

Do the means of the two groups differ significantly?

4. A group of students with a graduation index of 5 or better was compared with another group whose indices were 3.75 or below on a scale measuring attitude toward academic work. On the scale, the lower the score the more favorable the attitude and the higher the graduation index, the better the student.

	N	$\overline{X}$	s
Grad. Index 5.00+	15	34.2	10.0
Grad. Index 3.75–	14	45.2	12.5

Do the two means differ significantly?

5. Given $\overline{X}_1 = 10, N_1 = 10, \Sigma x_1^2 = 90, \overline{X}_2 = 13.1, N_2 = 15$, and $\Sigma x_2^2 = 146$. Compute and interpret t for these data.

6. Take the data of variable 1 in problem 5 above and set up the 95 percent and 99 percent confidence interval for this mean.

   95 percent interval

   99 percent interval

7. In a study the Σx^2 for the first variable is 104 and $N = 10$; for another group in the same study $\Sigma x^2 = 328$ and $N = 15$.
   Make an F test for these data and interpret your result.

8. Nine individuals were given a short psychomotor test. After a period of instruction, they were each given the same test again with the following results:

Ind.	Trial 1	Trial 2
1	18	24
2	10	10
3	8	12
4	16	18
5	7	11
6	11	15
7	4	8
8	20	22
9	16	18

   Is there a significant difference between the two means?

9. Given the following data:

$\overline{X}_1 = 60.4$	$r_{12} = .74$	$s_1 = 5.5$
$\overline{X}_2 = 66.3$	$N = 100$	$s_2 = 6.0$

Do these two means differ significantly?

10. A statistics class was given a test in which 34 students took the white form and 29 the yellow form. Actually the two tests were same except for the arrangement of the items. The mean on the white form was 28.9, $s = 6.94$; on the yellow form the mean was 30.4, and $s = 5.90$. Was there a difference in the performance of the students on these two forms?

chapter 13
TESTING DIFFERENCES BETWEEN PROPORTIONS

I. COMPLETION

When $p = .44$, $q =$ (1) ______. 1. ____________

If the parameter value of a proportion is .92, the sampling distribution is (2) ______. 2. ____________

To use the formula for the standard error of a proportion, Np or Nq, whichever is the smaller, should be at least (3) ______. 3. ____________

The sampling distribution of p is normal when p is closest to (4) ______. 4. ____________

p used in the formula for the standard error of the difference between two proportions is the proportion of individuals in (5) ______. In the case where 40 out of 80 in one group and 30 out of 90 in a second group agree to an item, p is (6) ______. This p is used to get a better estimate of the (7) ______. 5. ____________ 6. ____________ 7. ____________

Two proportions based on samples of 200 each are tested to see if they differ significantly. If $z = 2.64$, in respect to the H_0, we can say that (8) ______. 8. ____________

The value of the standard error of p is greatest when p equals (9) ______. 9. ____________

As the N upon which proportions are based increases, the size of the standard errors of the proportions (10) ______. 10. ____________

If the standard error of a proportion is .032, the standard error of the corresponding percentage is (11) ______. 11. ____________

II. PROBLEMS

1. In the course of a year, 70 patients were discharged from a mental hospital. Records were kept to see how many of these patients were readmitted to the hospital within one year after discharge. The patients were divided into short-term (hospitalized less than six months) and long-term (hospitalized six months or more). The results:

	Readmitted	Not Readmitted
Short-term	12	8
Long-term	20	30

Is length of hospitalization related to readmission? Test using difference between proportions.

2. The following data were obtained from an item analysis of test responses.

Item 1	Right	Wrong
High Group	82	18
Low Group	41	59

Item 2	Right	Wrong
High Group	38	62
Low Group	18	82.

Item 3	Right	Wrong
High Group	64	36
Low Group	58	42

Calculate a z for each, that is, test the significance of the difference between the proportions or percentages responding correctly to each item. Do your work under each set of data and interpret each z as to its significance.

3. Use the Lawshe-Baker nomograph in the text to the test the differences for each item by this method.

4. In a group of 50, 18 show a preference for Brand A and 32 for Brand B. In another group, 33 show a preference for Brand A and 50 for Brand B. Do the preferences of the two groups differ?

5. Given: 38 percent of a group of 100 individuals respond favorably to item 1 on an attitude scale and 42 percent of the same group respond favorably to item 2 on the same scale. Also, the correlation between the responses to items 1 and 2 is .54. Is there a significant difference between these two percentages?

6. On another attitude scale, item responses were tallied as shown:

		Item 1	
		No	Yes
Item 2	Yes	60	12
	No	20	18

Make a test for significance of the difference for these data using a formula for correlated data.

chapter 14
χ^2—CHI SQUARE

I. COMPLETION

Chi-square belongs to a group of statistical tests classified as (1) ______. When using such tests no assumption of (2) ______ is made in the (3) ______.

1. ______
2. ______
3. ______

To use chi-square, data must be in (4) ______.

4. ______

In a 2 X 2 table we have (5) ______ degree(s) of freedom. Such a table is referred to as a (6) ______ table.

5. ______
6. ______

Two major uses of chi-square are (7) ______ and (8) ______.

7. ______
8. ______

A contingency table with 4 rows and 5 columns has (9) ______ degrees of freedom.

9. ______

When the expected frequencies are small and the number of degrees of freedom is 1, chi-square should be corrected for (10) ______.

10. ______

With a *df* of 1, chi-square equals (11) ______.

11. ______

When testing to see if a distribution departs from normal (i.e., when testing goodness of fit), the curve constructed is based on the same (12) ______, the same (13) ______, and the same (14) ______ as the original data.

12. ______
13. ______
14. ______

When chi-square is applied in such a measure of goodness of fit, the number of degrees of freedom is equal to the (15) ______ minus (16) ______.

15. ______
16. ______

Before a chi-square problem is solved, the marginal cells of the observed frequencies must be (17) ______ to those of the expected frequencies.

17. ____________

The basic formula for the number of degrees of freedom in a chi-square problem is (18) ______ times (19) ______.

18. ____________

19. ____________

In a 4 X 1 contingency table there are (20) ______ degrees of freedom.

20. ____________

The categories of a contingency table must be (21) ______.

21. ____________

The smallest expected frequency that should be used when computing chi-square is (22) ______.

22. ____________

The contingency coefficient is a measure of relationship similar to (23) ______. With the latter data there are (24) ______ categories, whereas with the contingency coefficient there may be (25) ______.

23. ____________

24. ____________

25. ____________

A disadvantage of the contingency coefficient as a measure of relationship is that it does not have (26) ______ as an upper limit. Its upper limit is a function of the (27) ______.

26. ____________

27. ____________

The significance of the contingency coefficient is tested by (28) ______.

28. ____________

A related statistic, Cramér's statistic, differs from *C* in that its upper limit is (29) ______.

29. ____________

II. PROBLEMS

1. On a true-false test a student answers 40 out of 60 items correctly. Does this score of 40 differ from what would be expected by chance?
2. A die is tossed 300 times with the following results. Test to see if these observed frequencies differ from what would be expected by chance.

six-spot	five-spot	four-spot	three-spot	two-spot	one-spot
60	52	48	54	46	40

3. From a well-shuffled deck of regular playing cards 200 cards were drawn with the following results:

Red card:	115
Black card:	85
	200

Does this differ from what you would expect by chance?

4. In problem 1 of Chapter 10 there is a problem related to the tossing of coins. From Chapter 10 copy your observed frequencies and test to see if they differ from what would be expected by chance.

5. Patients in a mental hospital were given either tranquilizer pills or placebos in a study with the following results:

	Placebo	Regular Drug
Improved	8	12
Not improved	12	11

Test to see if the drug was effective.

6. Two groups of students respond to a Likert-type scale item as follows:

	Strongly Agree	Agree	No Opinion	Disagree	Strongly Disagree
Freshmen	12	32	10	23	23
Seniors	28	22	12	20	18

Do the responses of the two groups differ?

7. In another study with a similar scale the following data were obtained:

	Strongly Agree	Agree	No Opinion	Disagree	Strongly Disagree
Alumni	2	18	2	38	2
Seniors	6	24	3	12	4

Do these responses differ?

8. Eight dimes were tossed 256 times with the following results:

0 Heads	8 Tails	2
1 Head	7 Tails	12
2 Heads	6 Tails	30
3 Heads	5 Tails	52
4 Heads	4 Tails	68
5 Heads	3 Tails	54
6 Heads	2 Tails	27
7 Heads	1 Tail	10
8 Heads	0 Tails	1
		256

Do these results differ from what would be expected by chance?

9. The following data represent the responses of two groups to an attitudinal test item:

	Agree	Disagree
Males	70	30
Females	50	70

By the use of formula 14.4 in the text, test to see if the responses of the sexes differ on this item.

10. In working a chi-square problem the following observed and expected frequencies were obtained:

	O		E	
Group 1	10	21	10	16
Group 2	11	16	10	16
Group 3	9	15	10	16

Why should the work be stopped here?
Correct what is wrong with the above and find chi-square.

11. Given the data obtained on three groups of subjects:

	Pass	Fail
I	15	35
II	40	35
III	55	20

a. Compute the contingency coefficient for these data.

b. What is the upper limit of C for these data?

c. Compute Cramér's statistic for the above data.

chapter 15

AN INTRODUCTION TO THE ANALYSIS OF VARIANCE

I. COMPLETION

If there are six groups, (1) ______ separate *t* tests could be made. The analysis of variance makes it possible to test at once a general (2) ______ about all the group means.

Before we use the analysis of variance, we must have (3) ______ sampling, (4) ______ of variance among the subgroups, and (5) ______ of the samples. When we refer to homogeneity of variance, we mean that (6) ______.

The analysis of variance is said to be (7) ______. This means that certain (8) ______ may be violated and that no major effects on the outcomes will be produced.

In the single classification analysis of variance, the investigator is studying to see if there are differences between the (9) ______ among the (10) ______.

In the single classification analysis of variance there are three basic types of variance, (11) ______, (12) ______, and (13) ______. One of these is the measure of each of the separate scores about the (14) ______, another is the measure of the variability of the scores in each group about the (15) ______, and finally each of the individual (16) ______ varies about the (17) ______.

1. ______________
2. ______________
3. ______________
4. ______________
5. ______________
6. ______________
7. ______________
8. ______________
9. ______________
10. ______________
11. ______________
12. ______________
13. ______________
14. ______________
15. ______________
16. ______________
17. ______________

$\Sigma X^2 - (\Sigma X)^2/N$ results in the (18) ______. This, divided by the number of (19) ______, results in the variance or the (20) ______.

18. ______

19. ______

20. ______

When the within sum-of-squares is subtracted from the total sum-of-squares we obtain the (21) ______.
When the within sum-of-squares is divided by its number of degrees of freedom, we obtain the (22) ______, which may be referred to as the within (23) ______. Doing the same to the between sum-of-squares results in the (24) ______. Both of these are estimates of the (25) ______. The F test is applied to these two variances to determine if the two variances differ significantly. If the F is not significant, we conclude that (26) ______. In this case F is defined as the ratio of the (27) ______ to the (28) ______.

21. ______

22. ______

23. ______

24. ______

25. ______

26. ______

27. ______

28. ______

The F table is entered with the number of degrees of freedom for the (29) ______ variance on the abscissa and the (30) ______ variance on the ordinate.

29. ______

30. ______

Suppose that we have 4 groups of 8 cases each. The total number of degrees of freedom is (31) ______, the number of degrees freedom in each group is (32) ______, and *df* for the between variance is (33) ______.

31. ______

32. ______

33. ______

If an F is shown to be significant beyond the 1 percent level ($p < .001$), we determine where the actual differences are by use of (34) ______.
When there are only 2 groups, the value of F is equal to (35) ______.
In a two-way classification analysis of variance, the error variance is broken down into variances associated with the different (36) ______ in the study. In the two-way classification of ANOVA, the denominator of the F test is (37) ______. Any F less than (38) ______ is not significant. The F test is actually a (39) ______ test.

34. ______

35. ______

36. ______

37. ______

38. ______

39. ______

In a factorial design we have (40) ______ groups that have been measured after having been given varying amounts of a (an) (41) ______. In the interaction experiment the within sum-of-squares is broken down into various sums-of-squares for the (42) ______ under consideration plus the (43) ______ between factors.

40. ____________

41. ____________

42. ____________

43. ____________

II. PROBLEMS

1. Five individuals were given ten trials at a learning task with the following results:

	Individuals				
Trials	A	B	C	D	E
1	32	48	28	39	25
2	35	39	28	41	22
3	30	42	26	40	20
4	28	40	18	32	19
5	18	28	19	30	20
6	15	22	17	28	15
7	16	20	16	29	14
8	14	16	10	27	8
9	12	15	8	26	6
10	8	12	6	22	5

a. Apply the analysis test to see if the individuals differ in their performance.

b. Apply Scheffé's method to find out exactly where the differences are.

2. A, B, C, and D are different methods of teaching a skill. The scores represent the performance of 6 individuals in each of the 4 groups. Apply the analysis of variance to see if results obtained with the four methods differ.

Groups			
A	B	C	D
12	12	8	3
10	8	6	4
7	6	4	8
8	4	7	10
6	4	6	4
4	3	2	2

3. Below are the scores of two groups of individuals on a short test:

Group I	Group II
10	6
9	4
7	2
6	1
8	4
8	3
6	0
8	1
8	2
9	3
8	4
9	2
8	5

   a. By means of the analysis of variance determine if the means of the two groups differ significantly.

b. Using Fisher's formula (12.5) compute t for the same data.

c. Compare your F and t. What do you conclude?

4. An industrial psychologist built a short scale in studying accident proneness in a manufacturing plant. He administered his scale to two groups, 30 males and 30 females. In each group 15 had a record of no reported accidents and the other 15 had three or more reported accidents. By use of the analysis of variance see if his scale differentiated either by sex or number of accidents. Use the method of analysis of Table 15.5 in the text.

Males		Females	
No Accidents	Three or More Accidents	No Accidents	Three or More Accidents
8	12	6	10
12	8	4	6
16	18	8	8
4	14	4	12
6	13	3	14
6	8	7	16
8	6	10	7
2	16	12	16
7	5	5	3
8	10	3	10
5	16	7	12
6	14	6	14
7	12	4	13
3	8	6	11
5	10	2	8

5. Two groups, one of 40 male and one of 40 female college students, were each divided into four groups of 10. Each of the eight groups was then shown a different film depicting aspects of black-white relations. After the film was run an attitude scale was administered with the following results:

Males				Females			
1	2	3	4	1	2	3	4
10	14	13	16	10	14	10	18
8	12	12	12	14	13	10	16
6	8	10	10	12	12	10	15
4	4	9	9	12	11	9	14
4	4	9	9	8	10	9	13
4	4	7	7	6	9	7	10
2	2	4	7	4	7	7	10
2	2	4	6	2	6	6	10
2	1	4	5	2	2	4	10
1	3	2	5	1	3	5	9

a. Are the responses of the males and females different?
b. Did the different films affect the students' attitudes differently?
c. Did male and female students interact differently to the different films?

chapter 16

TESTING THE SIGNIFICANCE OF CORRELATION COEFFICIENTS

I. TRUE-FALSE

1.	If the parameter correlation coefficient is .85, the sampling distribution will be positively skewed.	1.	T	F
2.	The Pearson r ranges from -1 to +1.	2.	T	F
3.	When applying the null hypothesis to a Pearson r, we assume that the r does not differ from zero.	3.	T	F
4.	Any r above .90 is significant.	4.	T	F
5.	The sampling distribution of r is most likely to be normal when the parameter of r is .50.	5.	T	F
6.	The larger the sample, the greater the skew of the sampling distribution of r.	6.	T	F
7.	An r based on an N of 18 pairs would have 16 degrees of freedom.	7.	T	F
8.	An r of .20 based on 47 pairs of scores is significant at the 5 percent level.	8.	T	F
9.	An r of .40 based on 16 cases is not significant.	9.	T	F
10.	An r of .20 based on 92 cases is significant at the 1 percent level.	10.	T	F
11.	Fisher's Z statistic differs from the Pearson r in that sampling distribution of Z is normal throughout its range.	11.	T	F
12.	An r of .23 is equivalent to a Z of .23.	12.	T	F
13.	The difference between r's of .90 and .85 is larger than the difference between r's of 20 and 15.	13.	T	F
14.	We test the significance between two r's by testing the significance of the difference between two Z's.	14.	T	F
15.	When the difference between two Z's divided by the standard error of the difference between the two Z's is 2.61, we conclude that the difference between the two r's is significant at the 5 percent level.	15.	T	F
16.	In setting up the confidence interval for r, we use the standard error of r.	16.	T	F
17.	A point-biserial r is a Pearson product-moment r.	17.	T	F
18.	The statistical significance of the point-biserial r may be tested by using the table for testing the significance of the Pearson r.	18.	T	F

19. A biserial r is another type of Pearson r.	19.	T	F
20. When the point-biserial r and the biserial r are computed for the same data, the former is always larger.	20.	T	F
21. The standard errors of both the biserial and point-biserial are at a minimum when the point of dichotomy is .75.	21.	T	F
22. The point-biserial r is less reliable than the biserial r.	22.	T	F
23. The tetrachoric r is another Pearson product-moment r.	23.	T	F
24. The standard error of the tetrachoric r has been well worked out and is easy to use.	24.	T	F
25. Phi is a product-moment correlation coefficient.	25.	T	F
26. The significance of phi is tested using chi-square.	26.	T	F
27. Phi equals the square root of the quotient of chi-square over N.	27.	T	F
28. The F test used in evaluating the significance of eta is the ratio of the within variance to the between variance.	28.	T	F
29. Another name for mean-square is variance.	29.	T	F
30. When N is over 10, rho may be evaluated for significance in the same manner as the Pearson r.	30.	T	F
31. To evaluate the significance of W, the F test is used.	31.	T	F
32. When W is being evaluated, the larger the number of judges the smaller the size of W needed for significance.	32.	T	F
33. The most direct test for evaluating the significance of the phi coefficient is the F test.	33.	T	F
34. When an r of .80 and an r of .90 are averaged, the average r is .85.	34.	T	F
35. The average of two Z coefficients may be used as an average of the two Pearson r's.	35.	T	F

II. PROBLEMS

1. Given an N of 20 and an r of .60. By the use of an appropriate formula test the significance of this r.

By use of Appendix F in the text, check the significance of this *r*. How do the results compare?

2. Using Appendix G in the text, transform the following *r*'s to *Z*'s.

r	*Z*	*r*	*Z*	*r*	*Z*
.15		.65		.90	
.25		.80		.99	

3. Given:

$r_{xy} = .65$ $\quad r_{rs} = .75$
$N = 126$ $\quad N = 98$

By the use of appropriate formulas test the significance of the difference between these two *r*'s.

4. In a study based on an *N* of 150, a Pearson *r* of .62 was obtained. Set up the 99 percent confidence interval for this *r*.

5. In a study of 103 individuals three tests were taken that resulted in the following intercorrelations:

	1	2	3
1	–	.68	.74
2		–	.72
3			–

Test to see if the correlation between 1 and 2 (r_{12}) differs significantly from that between 1 and 3 (r_{13}).

6. Determine if the following point biserials are statistically significant.

r_{pb}	N	Significance
.44	27	
.44	42	
.27	18	
.31	82	
.60	10	

7. In the computation of r_b the following data were obtained: $N = 200$, $r_b = .50$, $p_1 = .33$, $p_2 = .67$. Is this r_b statistically significant?

8. An instructor used the phi coefficient as an index of discrimination in making an item analysis of test items. The analysis was based on 150 papers. Test the significance of each of the following phi coefficients that he obtained.

Item	Phi	Significance	Item	Phi	Significance
1	.42		5	.20	
2	.66		6	.05	
3	.12		7	.27	
4	.38		8	.19	

9. An eta coefficient was computed with data set up in 13 columns. The following sums were obtained:

$SS_t = 3841$
$SS_b = 1632$
$N = 100$

Test the significance of this correlation ratio.

10. By use of the appropriate technique test the significance of the following Spearman rank-order correlation coefficients.

rho	N	Significance
.40	18	
.37	37	
.30	28	
.40	102	
.13	80	

11. Four judges ranked 6 objects, the data producing a W of .90. Test the significance of this W.

    In a second case 10 judges ranked 8 objects. From these data a W of .21 was obtained. Is this W of statistical significance? Of practical significance?

12. The following r's were based on independent samples drawn from the same two variables.

    .88
    .78
    .60
    .84
    .92
    .70
    .80
    .90

    What is the average of these r's?

chapter 17

RELIABILITY, VALIDITY, AND ITEM ANALYSIS

I. COMPLETION

A test score is made up of two components, (1) _____ and (2) _____.

Systematic errors differ from random errors in that their mean is (3) _____.

The realibility of any test is inversely related to the (4) _____.

Test variance may be thought of as being made up of (5) _____ and (6) _____ variance, and the reliability of a test defined as that part of the variance that is (7) _____.

A measuring instrument that produces (8) _____ results is said to be (9) _____.

A test-retest reliability coefficient is referred to as a coefficient of (10) _____. The longer the period between testings, in general, the (11) _____ the reliability coefficient.

The correlation coefficient obtained between scores on two parallel tests is known as a coefficient of (12) _____.

With a test in which speed is an important factor, the most appropriate reliability coefficient is the (13)_____ coefficient.

In general, the (14) _____ a test, the more (15) _____ it is.

1. ____________
2. ____________
3. ____________
4. ____________
5. ____________
6. ____________
7. ____________
8. ____________
9. ____________
10. ____________
11. ____________
12. ____________
13. ____________
14. ____________
15. ____________

The reliability coefficient computed between the odd and even items of a test represents the reliability of a test (16) _____ the length of the original test. To obtain the reliability of a test of the original length the (17) _____ is applied.

16. __________

17. __________

The split-half and the Kuder-Richardson No. 20 reliability are both measures of the (18) _____ of a test. To obtain the latter one must have the (19) _____ of each item on the test.

18. __________

19. __________

The reliability of the well-made standardized test is typically above (20) _____.

20. __________

The reliability coefficient is related to the (21) _____ of the group tested, the greater the (22) _____ of the group, the higher the reliability coefficient. A measure not affected by this factor is the (23) _____.

21. __________

22. __________

23. __________

If the standard error of measurement of any test is 2, we can say that the chances are 2 out of 3 than an individual's (24) _____ score is not more than (25) _____ points from his (26) _____.

24. __________

25. __________

26. __________

When a test in constructed to cover adequately the material that it is supposed to cover, the test is said to have (27) _____ validity.

27. __________

Studying the psychological aspects of a trait measured by a test results in (28) _____ validity.

28. __________

When test scores are correlated with a (29) _____ we have (30) _____ validity. Research shows that the typical validity coefficient falls within the range (31) _____. This is caused to a great extent by the (32) _____ of the criterion.

29. __________

30. __________

31. __________

32. __________

The efficiency of prediction may be increased by the use of (33) _____. The statistic associated with errors of prediction is the (34) _____. When r is 1, the value of this statistic is (35) _____ and, when r is .00, the value of the statistic is the equivalent of (36) _____.

33. __________

34. __________

35. __________

36. __________

r^2 is called a coefficient of (37) _____ and k^2 a coefficient of (38) _____. As r increases, k (39) _____ but not at the (40) _____.

37. __________
38. __________
39. __________
40. __________

An index of forecasting efficiency of 8% means that in a specific prediction scheme we can predict (41) _____.

When we correct for attenuation in the criterion, we are correcting the criterion measures for (42) _____.

When a validity coefficient is attenuated, it is (43)_____.

41. __________
42. __________
43. __________

The ratio of the number of correct responses to a test item over the number reaching that item is a measure of the (44) _____ of the item. On a typical achievement test in order to have maximum discrimination throughout the range, the optimum difficulty level should average (45) _____ with items ranging from (46) _____ to (47) _____.

44. __________
45. __________
46. __________
47. __________

When one corrects test scores for guessing, one assumes that all errors are the result of (48) _____.

An item discriminates when it separates (49) _____ from (50) _____. Measures of discrimination such as the biserial and point-biserial r's are indices of the relationship between (51) _____ and (52) _____. An item that is responded to correctly by everyone has (53) _____ discrimination, and one with a negative index of discrimination is answered correctly by (54) _____.

48. __________
49. __________
50. __________
51. __________
52. __________
53. __________
54. __________

Item analyses should be based on (55) _____ and any analysis made with groups of say 25 should be expected to produce (56) _____ results.

55. __________
56. __________

For the distractors of multiple-choice items to be effective ones, they must be more attractive to the (57) _____ students.

Lambda gives an indication of the (58) _____ in the probability of error in predicting X from Y. In this way it differs drastically from (59) _____.

57. __________
58. __________
59. __________

II. PROBLEMS

1. Below are the responses of 30 students to 16 items.

Student	Item Number 1	2	3	4	5	6	7	8	9	10	11	12	13	14	15	16
A	X	X	X	X	X	X	X	X	O	X	X	X	X	X	X	X
B	X	X	X	X	O	O	O	X	X	X	X	X	X	O	X	X
C	X	O	X	X	X	X	X	O	X	X	O	X	O	X	X	X
D	X	X	X	X	X	X	X	X	X	X	O	X	O	O	O	O
E	X	X	X	X	X	X	O	O	X	O	X	X	O	X	O	X
F	X	X	X	X	X	O	X	X	O	X	O	O	O	O	X	X
G	X	X	X	X	X	X	O	O	O	X	X	O	O	X	X	O
H	X	X	X	X	X	X	X	X	X	X	O	X	X	O	X	O
I	X	X	O	X	X	X	X	X	O	X	X	O	O	X	O	X
J	X	X	X	X	X	X	O	X	X	O	X	X	O	O	X	O
K	X	X	X	X	O	O	O	X	X	O	X	O	O	O	O	X
L	O	O	X	X	X	X	X	X	X	O	O	O	O	O	X	O
M	X	X	X	O	X	X	X	X	X	X	O	X	X	O	O	O
N	X	X	X	X	X	X	X	O	O	X	X	O	X	X	O	O
O	X	X	X	X	O	X	O	O	X	X	O	O	O	O	O	X
P	X	X	X	X	X	X	O	O	X	X	X	X	X	X	O	O
Q	X	X	X	X	X	X	O	O	O	O	X	O	X	X	O	O
R	X	O	X	X	O	X	X	O	O	O	X	X	O	O	O	X
S	X	O	O	X	X	O	O	X	X	O	O	O	X	X	X	O
T	X	O	X	X	O	O	X	X	X	O	X	X	O	O	X	X
U	X	X	O	X	X	O	X	O	X	O	X	X	X	X	O	O
V	X	X	X	X	O	O	O	X	O	X	O	O	O	X	O	X
W	X	X	O	O	X	X	O	O	X	X	O	O	O	O	O	O
X	X	O	O	X	X	X	O	O	O	O	X	O	O	O	X	X
Y	X	X	X	X	O	X	X	O	O	X	O	O	O	O	X	O
Z	X	X	X	X	O	O	O	O	X	O	O	O	O	O	O	O
AA	X	X	X	O	O	O	O	X	X	O	X	X	X	X	O	O
BB	X	X	X	X	O	O	O	X	O	O	O	X	O	X	O	O
CC	X	O	O	X	X	O	X	O	O	X	X	O	X	X	X	O
DD	O	O	O	X	X	O	O	O	O	O	X	O	X	X	O	O

X = item answered correctly
O = item answered incorrectly

a. For the preceding data obtain an odd and an even score for each individual. Use these odd and even scores to compute the reliability of the test.

b. Compute the reliability of the preceding test using Kuder-Richardson formula 20.

c. Using the results obtained in a above, determine the standard error of measurement of this test.

d. For an individual with a score of 10 on this test, interpret the standard error of measurement found in c above.

2a. A 20-item test has a reliability coefficient of .70. To this test are added 60 well-made and similar items. What is the reliability of the new test?

b. A 210-item test takes two hours to administer. It is decided to shorten the test to a 70-item test so that it may be administered in 40 minutes. The reliability of the original test was .94. What might you expect the reliability of the shortened test to be when similarly used?

c. A 10-item test has a reliability coefficient of .50. How many well-made and similar items will have to be added to increase the reliability to .90?

3a. Suppose that a certain test of academic ability has a reliability coefficient of .90 and a validity coefficient of .50 when correlated against freshmen grades at a certain college. The standard deviation of the test is 100 and that of freshmen grades .8. The criterion has a reliability of .60. Calculate the standard error of estimate in predicting this criterion measure.

b. For the above data a regression line has been constructed and from this regression line for a given score on the test a predicted Y value of 4.8 is obtained. How is the standard error of estimate obtained in a above related to this predicted index of 4.8?

c. Calculate and interpret the index of forecasting efficiency for these data.

4. Given that r_{xy} is .60, where X is a test and Y a criterion. This coefficient is squared.

a. What is it now called?

b. What does it mean?

5. Use the data of problem 3a above.

a. Correct for attenuation involving both test scores and criterion measures.

b. Correct for attenuation using only the criterion scores.

c. What is being done and when is this justified?

6. The following data were obtained from an item analysis:

Item	Number Correct Upper 27%	Number Correct Lower 27%	Percent Correct Upper 27%	Percent Correct Lower 27%	Difficulty	Flanagan *r*
1	82	28				
2	62	36				
3	32	2				
4	48	48				
5	20	2				
6	84	14				
7	38	62				
8	72	48				
9	68	22				
10	98	96				

The upper and lower 27 percent are each made up of 100 papers. Complete the above table. Obtain the Flanagan *r*'s from Fig. 17.1 in the text.

a. Which of the above items would be considered good? Why?

b. Which of the above items would be considered poor or worthless? Why?

7. Using Fig. 17.2 in the text, obtain a phi coefficient for the 10 items presented in problem 6.

Item	Phi	Item	Phi	Item	Phi
1		5		8	
2		6		9	
3		7		10	
4					

8. Test the significance of each Flanagan r obtained in problem 6 and each phi coefficient obtained in problem 7.

   What can you say about the two methods of determining whether or not an item discriminates?

chapter 18
DISTRIBUTION-FREE STATISTICAL TESTS

I. SHORT ANSWER

Each of the terms listed below pertains more to distribution-free (DF) statistics than it does to parametric (P), or vice versa. Mark the category into which you think each is best classified.

1. Nominal data	1. DF	P
2. Statistical power	2. DF	P
3. Simplicity of derivation	3. DF	P
4. Large samples	4. DF	P
5. Greater number of applications	5. DF	P
6. Time involved in computing	6. DF	P
7. More economical in time and energy	7. DF	P
8. Ordinal data	8. DF	P
9. *N* of 10	9. DF	P
10. Few assumptions	10. DF	P
11. Ratio or interval scale	11. DF	P
12. Normal distribution	12. DF	P

II. COMPLETION

The sign test is to be used when data are (1) ______. The basic assumption involved with this test is that the data are (2) ______. In this test we are concerned with changes and we assume that the (3) ______ change is zero. If the null hypothesis is true, we would expect to find an equal number of (4) ______ and (5) ______ changes. The extent of the departure from this 50-50 expectancy may lead to the rejection of (6) ______ if there is a large enough number of one sign. When *N* is less than 10, we obtain our probabilities from the (7) ______. This test is not (8) ______.

1. ______________
2. ______________
3. ______________
4. ______________
5. ______________
6. ______________
7. ______________
8. ______________

Wilcoxon's test is a test for (9) _____ data. In this test the difference between each pair of scores is (10) _____ and then these are placed into separate columns depending on the sign of the rank. If there is no difference in the samples, the sum of the plus and minus columns of ranks would be (11) _____ and the null hypothesis (12) _____. This test is (13) _____ than the sign test.

9. ________
10. ________
11. ________
12. ________
13. ________

The median test is for (14) _____ data. The rationale behind this test is that if the null hypothesis is true, there would be an (15) _____ number of cases above and below the (16) _____ in each category. The median test is completed by the use of (17) _____.

14. ________
15. ________
16. ________
17. ________

The Mann-Whitney U test is also based on (18) _____, the scores of the two distributions studied being all ranked together. The ranks are summed and the null hypothesis is actually rejected when the sums depart significantly from (19) _____.

18. ________
19. ________

If, in a runs test, the two samples are from the same population, the number of runs expected would be (20) _____; high, medium, and low scores being found about in the same numbers in each (21) _____. In a runs test ties are handled by (22) _____.

20. ________
21. ________
22. ________

The most powerful distribution-free test studied is the (23) _____.

23. ________

III. PROBLEMS

1.

X	Y	$X - Y$
26	18	
32	24	
16	20	
19	17	
12	6	
15	11	
22	24	
18	12	
16	12	
26	24	
27	21	
24	20	

Test the above data for a significant difference using the sign test. Since the number of pairs is greater than 10, the results should be evaluated using the formula for the mean and standard deviation of the binomial.

2.

X	Y	$X - Y$
6	8	
7	12	
15	18	
16	22	
8	6	
12	12	
10	18	
7	12	
13	17	

Apply the sign test to the above data.

3.

X	Y
26	18
32	24
16	20
19	17
12	6
15	11
22	24
18	12
16	12
26	24
27	21
24	20

The data above are the same as those in problem 1. Apply Wilcoxon's matched-pairs signed-ranks test to the above data. Compare your results with those obtained in problem 1.

4. Scores on a short vocabulary test.

Group I	Group II
12	24
18	18
26	26
11	19
18	19
10	10
8	12
10	18
13	22
12	24
22	
17	

Apply the median test to the above and interpret your results.

5. Apply the Mann-Whitney U test to the data in problem 4 which have been recopied below.

Group I	Group II
12	24
18	18
26	26
11	19
18	19
10	10
8	12
10	18
13	22
12	24
22	
17	

Interpret your results.

6. Below are scores of three different diagnostic groups on the arithmetic scale of the Wechsler-Bellevue.

A	B	C
12	6	10
14	4	4
16	3	6
12	5	6
10	8	7
8	12	6
	8	4
	7	2

Test for differences among the three groups, using the median test.

7. Below are the scores of a group of boys and girls on the clerical scale of an interest inventory.

Boys	Girls
10	17
12	22
13	18
7	20
13	16
3	12
18	20
5	18
7	16
11	6
12	18
	22
	14
	15
	21
	19
	6
	17
	18
	16
	20

Apply the Mann-Whitney U test to the above data and interpret your results.

8. Below are scores of two different samples:

A	B
32	28
25	26
14	17
12	16
13	17
27	26
11	25
30	27
17	19
20	31

Test the hypothesis of no difference using the Wald-Wolfowitz runs test.

9. Three groups are tested with a short-digit symbol test with the following results:

A	B	C
12	19	20
16	2	10
18	11	12
7	3	2
18	4	5
8	8	7
4		9
		13

Test for differences among the three groups using the Kruskal-Wallis H test and interpret your results.

Answers

CHAPTER 1

I. True-False

1. T	8. T	15. T	22. F
2. F	9. T	16. T	23. T
3. T	10. T	17. T	24. T
4. T	11. T	18. T	25. T
5. F	12. T	19. T	26. F
6. T	13. F	20. F	27. T
7. T	14. T	21. F	

CHAPTER 2

I. Completion

1. positive
2. negative
3. zero
4. subtrahend
5. add
6. reciprocal
7. Appendix A
8. smaller
9. negative
10. coefficient
11. variable concerned
12. exponent
13. whole
14. 1
15. 100
16. 100
17. discrete
18. number of children in a family
19. number of words typed per minute
20. number of deaths caused by cancer
21. continuous
22. nominal
23. ordinal
24. interval
25. a household thermometer
26. ratio
27. equal units of measurement
28. absolute zero
29. nominal
30. ordinal
31. interval
32. ratio
33. ordinal
34. so many centimeters taller
35. so many times taller
36. statistics
37. parameters
38. universe
39. random
40. sum of
41. is greater than
42. does not equal
43. population
44. sample
45. yes
46. a group of graduating seniors could be a population from which we take a sample, *or* the same group (or an equivalent group) could be a sample of graduating seniors from a group of colleges and universities

II. Exercises

1. 611
2. 15.2
3. -4.4
4. -3.13

Chapter 2: II. *Continued*

5. -4.2163
6. .0014
7. .64
8. .21
9. .37
10. .93
11. .000000207936
12. .001126
13. .0336
14. 69.7
15. 2.2
16. .03
17. .00947
18. .096
19. 256
20. 40
21. $x^3 y^4 z$
22. fx^2
23. 16
24. 193.6
25. .064
26. 90
27. .5625
28. 0
29. 1.1
30. .1
31. 109.8
32. 109.8
33. .1
34. 4
35. 3

CHAPTER 3

I. Completion

1. meaningful
2. range
3. range
4. 71
5. 10-20
6. 15
7. lower limit
8. midpoint
9. upper limit
10. 2-3 or 3-4
11. ordinates
12. abscissa
13. class interval
14. midpoints
15. percentages
16. reciprocal
17. frequency
18. 100
19. negative
20. leptokurtic
21. platykurtic
22. unimodal
23. zero
24. bilaterally symmetrical
25. meso-
26. leptokurtic
27. true limits
28. pie
29. bar
30. cumulative proportions or percentages
31. upper limits
32. point
33. 56%
34. median
35. C_{50}
36. Q_2
37. C_{25}
38. C_{70}
39. rectangular
40. little meaning
41. centile rank
42. S shaped
43. centiles
44. centile ranks

	Exact Limits	Interval Size
51.	-.5 − 3.5	4
52.	7.5 − 10.5	3
53.	79.5 − 89.5	10
54.	-5.5 − 5.5	11
55.	.495 − .755	.26

56. 1
57. 3
58. 10 preferred
59. .4, .5, .6, .7
60. 3, 4, 5

II. Problems

6. a. 82.5
   b. 67
   c. 56.2
   d. 55.5
   e. 38.2

9. Median, 674.1 Q_1, 643 C_{72}, 698.2 D_8, 710.4
   Q_3,702.4 D_1, 608.1 C_{95}, 745.3 D_2, 634.3

11. 760–99   666–43   575-3   470–0

CHAPTER 4

I. Completion

1. adding
2. number of scores
3. deviations
4. zero
5. deviation
6. x
7. sum of the squares
8. smaller
9. weighted or grand
10. mean
11. number of cases
12. has the same constant subtracted from it
13. divided by the same constant
14. point
15. 50%
16. mode
17. 74.5
18. normal
19. skewed
20. symmetrical
21. left-hand tail
22. negatively skewed
23. reliability
24. it varies considerably from sample to sample
25. median
26. mean
27. errors of grouping
28. trivial
29. midpoint
30. errors of grouping
31. mean
32. sensitive
33. mean
34. median
35. median
36. median
37. median
38. mean
39. mean
40. median
41. mode
42. median

II. Problems

1. $\overline{X} = 5.3$
2. a. $\overline{X} = 12.8$
   b. median = 13.5; mode = 16
3. $\overline{X} = 76.7$
   median = 80.3
   mode = 84.5
4. a. 17.8
   b. 37.8
   c. 8.9
5. 76.2
6. a. 12
   b. 11.3
   d. 10.7

CHAPTER 5

I. Completion

1. unreliability
2. deviation
3. mean
4. zero
5. less
6. average deviation
7. sum of the squares
8. variance
9. mean-square
10. 68
11. 95
12. 99.74
13. equal
14. 6
15. 4.1
16. 12
17. 144
18. standard deviation
19. variance
20. variance
21. quartile deviation
22. semi-interquartile range
23. 9.95
24. interquartile range
25. 50
26. s
27. mean
28. Q
29. normal
30. skewed or departs from normal in some other way
31. heterogeneous
32. homogeneity
33. 53.3
34. 8066
35. zero
36. variance
37. $\Sigma x^3/N$

Chapter 5: Completion *Continued*

38. second
39. third
40. zero
41. positive
42. negative
43. second
44. fourth
45. mesokurtic
46. leptokurtic
47. platykurtic

II. Problems

1. $\bar{X} = 54.5$
   $s = 20.8$
2. $\bar{X} = 47.8$
   $s = 11$
3.

	$\bar{X}$	AD	s
a.	19.8	6.0	6.7
b.	19.4	6.4	7.8
c.	67.1	9.9	11.6

4. a. 6.7
   b. 13.4
5. a. $\bar{X}_t = 149.7$
   b. $s_t = 13.1$
6. $Q = 7.5$
7. $Q = 9.5$
8. a. $m_1 = 0$ $m_3 = -356.1$
   $m_2 = 85.2$ $m_4 = 22037.4$
   b. $g_1 = -.453$ $g_2 = .07$

CHAPTER 6

I. Completion

1. deviation
2. standard deviation units; sigma units
3. zero
4. 1
5. mean
6. standard deviation
7. either positive or negative
8. 6 sigmas
9. new sigma
10. new mean
11. 660
12. 66
13. 23
14. different tests
15. standard scores
16. positively skewed
17. an infinite number
18. asymptotic
19. zero
20. meso-
21. mean
22. one
23. unity
24. frequencies or cases
25. centile rank
26. 20
27. 68
28. 95
29. 99.74
30. .0026
31. .9974
32. same mean
33. same standard deviation
34. same number of cases
35. No
36. Yes
37. No
38. Yes
39. No
40. No

II. Problems

1.

	z score	Transformed Score
a.	0	50
b.	1.00	60
c.	-1.00	40
d.	2.50	75
e.	-2.50	25
f.	.625	56

2.

a.			b. Average	Rank
59	60	83	67.3	4
56	69	80	68.3	3
41	42	48	43.7	8
52	46	55	51.0	6
63	61	72	65.3	5
70	68	72	70.0	2
72	71	77	73.3	1
74	41	32	35.7	10
39	35	51	41.7	9
44	54	43	47.0	7

3. a. 59.7
   b. B+ or A-

Chapter 6: Problems *Continued*

4.

	Area	Centile
a.	.1554	C_{66}
b.	.4713	C_{97}
c.	.2486	C_{25}
d.	.3413	C_{84}
e.	.4951	$C_{99.5}$

5. a. 416
   b. 40
   c. 548
   d. 4
6. a. 153
   b. 466
   c. 22
7. $\bar{X} = 73.6; s = 10.6$

CHAPTER 7

I. Completion

1. bivariate
2. relationship
3. causality
4. equal to
5. negative; inverse
6. no
7. ± 1
8. .00
9. fall along a straight line
10. are scattered
11. curvilinear
12. linear regression
13. means
14. straight line
15. negative
16. one
17. homoscedasticity
18. *z* scores
19. mean *z* score product
20. covariance
21. variance
22. variance
23. covariance
24. scatterplot
25. linearity
26. relationship
27. smaller
28. range
29. number
30. underestimate
31. eta; correlation ratio
32. related to a third variable
33. chance
34. consistency
35. .90
36. validity
37. criterion
38. smaller
39. .40-.60
40. chance
41. true
42. equal
43. prediction
44. high positive
45. medium positive
46. low positive
47. none
48. medium negative
49. high negative
50. curvilinear
51. about zero
52. about zero
53. high positive
54. about zero
55. high positive
56. about zero
57. moderate negative

II. Problems

1. a. $r = .88$
   b. $\bar{X} = 44.6$
   $\bar{Y} = 34.3$
   $s_x = 10.3$
   $s_y = 10.4$
2. $r = .84$
3. $r = .84$
4. $r = .84$
5. b. $r = .82$
   $\bar{X}_{AS} = 14.0$
   $\bar{X}_{BS} = 15.0$
   $s_{AS} = 4.1$
   $s_{BS} = 4.6$

CHAPTER 8

I. Completion

1. dichotomy
2. continuous
3. true dichotomy
4. Pearson *r*
5. proportion of total group responding correctly

Chapter 8: Completion *Continued*

6. 1 - *p*
7. abacs
8. forced dichotomy
9. ordinate in normal curve cutting off *p*
10. biserial
11. point biserial
12. arbitrary
13. phi
14. true dichotomies
15. groups are equal size
16. tetrachoric
17. table
18. *ad*/*bc*
19. reliable
20. biserial
21. tetrachoric
22. underestimate
23. correlation ratio
24. eta
25. sum-of-squares
26. total
27. there is a linear relationship
28. data depart from linearity
29. linearity
30. eta has no sign
31. third
32. two other variables
33. combined effects
34. ranks
35. .89
36. small; less than 30
37. *W*
38. point biserial
39. rho
40. phi
41. *W*
42. partial *r*
43. *R*
44. eta
45. point biserial
46. Pearson *r*
47. tetrachoric
48. eta

II. Problems

1. $r_{pb} = .64$
2. $r_{pb} = .54$
3. $r_b = .675$
4. phi = .20
5. phi = −.10
6. For problem 4, $r_t = .31$
   For problem 5, $r_t = -.16$
7. Partial $r = .05$
8. $R = .77$
9. eta = .90
10. rho = .88
11. $T = .76$
12. $W = .92$

CHAPTER 9

I. Completion

1. linear
2. independent
3. dependent
4. intercept
5. slope
6. *Y*
7. *X*
8. positive
9. negative
10. negative
11. error of prediction
12. sum of the squares
13. minimum
14. line of best fit
15. two *b* and two *a*
16. *Y*
17. *X*
18. *X*
19. *Y*
20. sum of the cross-products
21. sum of the squares
22. r^2
23. $\overline{X}$
24. $\overline{Y}$
25. .60
26. standard deviations
27. *r*
28. 70
29. 66.5
30. 65
31. 75
32. 48
33. error
34. *X* axis
35. $\overline{Y}$
36. Y'
37. $\overline{Y}$
38. *s*
39. zero
40. variability of scores
41. regression line
42. best fit
43. standard deviation

Chapter 9: Problems *Continued*

44. ±1 s_{yx} from the regression line
45. smaller
46. homoscedasticity
47. no
48. closer
49. criterion
50. constant
51. several
52. criterion
53. predictors
54. up to 4-5 predictor

II. Problems

1. a. 27
   b. -12
   c. 14
   d. -11.5
2. a. .924
   b. $Y' = -.56 + .90X$
   c. $X' = 1.09 + .95Y$
   d. .46
   e. $r^2 = b_{yx} b_{xy}$
3. a. $Y' = 4.35$
   b. $s_{yx} = .61$
4. a. $.393X - 4.9$
   b. 2.2
   c. 14

CHAPTER 10

I. Completion

1. certainty
2. 10–1
3. 1/8
4. sum
5. separate probabilities
6. independent
7. .00046
8. normal curve
9. discrete
10. binomial
11. .5
12. 1
13. *np*
14. mean
15. variance
16. 100
17. 24
18. $\frac{18.5 - 12}{3}$

II. Short Answer

1. a. .3
   b. .9
   c. 1.00
   d. .06
   e. .027
   f. .6
   g. .018
2. a. .0192
   b. .0769
   c. .25
   d. .058
   e. .25
   f. .265
   g. .0312
   h. .001953
   i. .055
   j. .044
   k. .5
   l. .0278
   m. .50

III. Problems

2. a. $p = .008$
   b. $p = .0087$
3. a. $p = .0161$
   b. $p = .0193$
   c. $p = .017$
   d. $p = .0217$
4. a. $m = 20$
   b. $p = .0071$
   c. $p = .0037$
5. $z = 1.37; p > .05$

CHAPTER 11

I. Completion

1. inference
2. nonprobability
3. probability
4. accidental
5. quota
6. convenient, economical
7. random
8. equal
9. biased
10. biased sample
11. stratified random
12. cluster
13. population

Chapter 11: Completion *Continued*

14. defined
15. table of random numbers
16. sampling
17. standard error of the mean
18. s
19. $\sqrt{N - 1}$
20. normal
21. 2 out of 3
22. 67-73
23. reliability
24. inversely
25. directly
26. unbiased
27. biased
28. correct for bias
29. standard error
30. 25%
31. reliable
32. $1 - p$
33. one
34. standard error of the mean
35. 99% confidence
36. 99%
37. population mean
38. wider
39. inferences
40. population
41. biased
42. no

II. Problems

3. a. $s_{\bar{x}_1} = .917$
   $s_{\bar{x}_2} = .889$

   b.

$s_{\bar{x}_1}$	$s_{\bar{x}_2}$
95% interval	95% interval
84.2-87.8	80.3-83.7
99% interval	99% interval
83.6-88.4	79.7-84.3

4. $s_{mdn_1} = 1.14$
   $s_{mdn_2} = 1.11$
5. a. $p = .1587$
   b. $p = .0062$

CHAPTER 12

I. Completion

1. statistic
2. null hypothesis
3. alternate hypothesis
4. alternate hypothesis
5. alpha level
6. II
7. I
8. I
9. 1.645
10. 2.33
11. 1.96
12. 2.58
13. 99 in 100
14. three
15. N is large
16. 6
17. 46
18. better
19. population
20. a deviation
21. standard deviation
22. homogeneity
23. F test
24. greater variance
25. lesser variance
26. random sampling
27. 1.42
28. accepted
29. robust

II. Problems

1. no; $z = 1.87; p > .05$
2. no; $z = 1.50; p > .05$
3. yes; $t = 3.41; p < .001$
4. $t = 2.53; p < .05$
5. $t = 2.37; p < .05$
6. 95% confidence interval:
   7.4–12.6
   99% confidence interval:
   6.8 – 13.2
7. $F = 2.10; p > .05$
8. $t = 5.30; p < .001$
9. $z = 14.08; p < .001$
10. no; $t = .94, p > .05$

CHAPTER 13

I. Completion

1. .56
2. negatively skewed
3. 5
4. .50
5. the two groups combined
6. .41
7. parameter proportion
8. it is rejected at the 1% level
9. .50
10. decrease
11. 3.2

II. Problems

1. $t = 1.52; p > .05$
2. Item 1: $z = 5.96; p < .001$
   Item 2: $z = 3.15; p < .01$
   Item 3: $z = .87; p < .05$
4. $z = -44; p > .05$
5. $z = .85; p > .05$
6. $t = 4.76; p < .01$

CHAPTER 14

I. Completion

1. distribution free
2. normality
3. population
4. frequencies
5. 1
6. contingency
7. testing goodness of fit
8. testing independence
9. 12
10. lack of continuity
11. z^2
12. mean
13. standard deviation
14. number of cases
15. number of intervals
16. 3
17. equal
18. rows minus 1
19. columns minus 1
20. 3
21. mutually exclusive
22. 5
23. phi
24. two
25. three or more
26. 1
27. number of categories
28. chi-square
29. 1

II. Problems

1. $X^2 = 6.67; p < .01$
2. $X^2 = 4.80; p > .05$
3. $X^2 = 4.50; p < .05$
5. $X^2 = .626; p > .05$
6. $X^2 = 9.254; p > .05$
7. $X^2 = 10.617; p < .01$
8. $X^2 = 3.815; p > .05$
9. $X^2 = 17.66; p < .001$
10. $X^2 = .437; p > .05$
11. a. $C = .321$
    b. .712
    c. .26

CHAPTER 15

I. Completion

1. 15
2. hypothesis
3. random
4. homogeneity
5. independence
6. variances do not differ significantly
7. robust
8. assumptions
9. means
10. groups
11. total
12. between
13. within
14. grand mean
15. group mean
16. means
17. grand mean
18. sum of the squares
19. degrees of freedom
20. mean-square
21. between sum-of-squares
22. mean-square
23. variance
24. between variance

Chapter 15: Completion *Continued*

25. population variance
26. null hypothesis stands
27. between variance
28. within variance
29. greater
30. lesser
31. 31
32. 7
33. 3
34. Scheffé test
35. t^2
36. columns and rows
37. mean-square for error
38. 1
39. robust
40. two
41. intervening variable
42. factors
43. interaction

II. Problems

1. a. $F = 5.67$
   b. C differs from D–5% level
   D differs from E–1% level
2. $F = .99; p > .05$
3. a. $F = 80.3; p < .001$
   b. $t = 8.94; p < .001$
   c. $F = t^2; 80.2 = 80.1$
4. $F = 1.32; p > .05$
   $F = 9.99; p < .01$
5. a. $F = 9.90; p < .01$
   b. $F = 6.16; p < .01$
   c. $F = .92; p > .05$

CHAPTER 16

I. True-False

1. F
2. T
3. T
4. F
5. T
6. F
7. T
8. F
9. T
10. F
11. T
12. T
13. T
14. T
15. F
16. F
17. T
18. T
19. F
20. F
21. F
22. F
23. F
24. F
25. T
26. T
27. T
28. F
29. T
30. T
31. F
32. T
33. F
34. F
35. T

II. Problems

1. $t = 3.18; p < .01$
2. | | | |
   |---|---|---|
   | .15 | .775 | 1.472 |
   | .255 | 1.099 | 2.647 |
3. $z = 1.29; p > .05$
4. 47 – 74
5. $z = 1.26; p > .05$
6. $p < .05$
   $p < .01$
   $p > .05$
   $p < .01$
   $p > .05$
7. $z = 5.44; p < .001$
8. 1. $p < .01$
   2. $p < .01$
   3. $p > .05$
   4. $p < .01$
   5. $p < .05$
   6. $p > .05$
   7. $p < .01$
   8. $p < .05$
9. $F = 5.35; p < .01$
10. $p > .05$
    $p < .05$
    $p > .05$
    $p < .001$
    $p > .05$
11. $p < .01$
    $p < .05$
12. .825

CHAPTER 17

I. Completion

1. true
2. error
3. not zero
4. size of the error component
5. true
6. error
7. true
8. consistent
9. reliable
10. stability
11. lower
12. equivalence
13. equivalence
14. longer
15. reliable
16. ½
17. Spearman-Brown formula
18. internal consistency
19. difficulty
20. .90
21. range of talent
22. range
23. standard error of measurement
24. obtained
25. 2
26. true score
27. content
28. construct
29. criterion
30. criterion related
31. .40-.60
32. deficiency and unreliability
33. several predictors
34. standard error of estimate
35. zero
36. standard deviation of the y distribution
37. determination
38. nondetermination
39. decreases
40. same rate
41. 8% better than chance
42. lack of reliability
43. lowered
44. difficulty
45. .50
46. very easy
47. very difficult
48. chance
49. good; high scoring
50. poor; low scoring
51. responses to an item
52. total test scores
53. zero
54. more of the high scorers than the low
55. large samples
56. unreliable
57. poorer, low-scoring
58. reduction
59. phi coefficient

II. Problems

1. a. $r = .53$
   b. $KR\text{-}20 = .36$
   c. 1.57
2. a. $r = .90$
   b. $r = .84$
   c. 80 items
3. a. .693
   c. 13.5%
5. a. .68
   b. .645
6.

Difficulty		Flanagan r
1.	55	.54 $p < .001$
2.	49	.27 $p < .02$
3.	17	.56 $p < .001$
4.	48	.00 $p > .05$
5.	11	.50 $p < .001$
6.	49	.68 $p < .001$
7.	50	-.25 $p < .02$
8.	60	.26 $p < .01$
9.	45	.47 $p < .001$
10.	97	.12 $p > .05$

7.

	Phi	Significance
1.	.54	$p < .001$
2.	.26	$p < .01$
3.	.40	$p < .001$
4.	.00	$p > .05$
5.	.30	$p < .01$
6.	.70	$p < .001$
7.	-.25	$p < .05$
8.	.25	$p < .05$
9.	.45	$p < .001$
10.	.10	$p > .05$

CHAPTER 18

I. Short Answer

1. DF
2. P
3. DF
4. P
5. DF
6. DF
7. DF
8. DF
9. DF
10. DF
11. P
12. P

II. Completion

1. correlated
2. continuous
3. median
4. plus
5. minus
6. H_0
7. binomial distribution
8. powerful
9. correlated
10. ranked
11. zero
12. would stand
13. more powerful
14. uncorrelated
15. equal
16. median
17. chi-square
18. ranks
19. the mean rank
20. large
21. sample
22. putting tied scores in all possible combinations
23. Kruskal-Wallis H test

III. Problems

1. $z = 2.02$; $p = .0404$
2. $p = .07$
3. $T = -7.5$
4. $X^2 = 4.71$; $p < .05$
5. $U_2 = 31$; $p > .05$
6. $X^2 = 11.00$; $p < .02$
7. $z = 3.35$; $p < .01$
8. $z = .03$; $p > .05$
9. $H = 1.61$; $p > .05$

Y-Axis	0	1	2	3	4	5	6	7	8	9	10	11	12	13	14	15	16	17	18	19	20	fy	y'	fy'	fy'^2	$x'y'$
24																							24			
23																							23			
22																							22			
21																							21			
20																							20			
19																							19			
18																							18			
17																							17			
16																							16			
15																							15			
14																							14			
13																							13			
12																							12			
11																							11			
10																							10			
9																							9			
8																							8			
7																							7			
6																							6			
5																							5			
4																							4			
3																							3			
2																							2			
1																							1			
0																							0			
fx																						N	$\Sigma=$			
x'	0	1	2	3	4	5	6	7	8	9	10	11	12	13	14	15	16	17	18	19	20					
fx'																						$\Sigma=$				
fx'^2																						$\Sigma=$				

X-Axis

Y-Axis	0	1	2	3	4	5	6	7	8	9	10	11	12	13	14	15	16	17	18	19	20	fy	y'	fy'	fy'^2	x'y'
24																							24			
23																							23			
22																							22			
21																							21			
20																							20			
19																							19			
18																							18			
17																							17			
16																							16			
15																							15			
14																							14			
13																							13			
12																							12			
11																							11			
10																							10			
9																							9			
8																							8			
7																							7			
6																							6			
5																							5			
4																							4			
3																							3			
2																							2			
1																							1			
0																							0			
fx																						N	Σ=			
x'	0	1	2	3	4	5	6	7	8	9	10	11	12	13	14	15	16	17	18	19	20					
fx'																						Σ=				
fx'^2																						Σ=				

Y-Axis	0	1	2	3	4	5	6	7	8	9	10	11	12	13	14	15	16	17	18	19	20	fy	y'	fy'	fy'^2	$x'y'$
24																							24			
23																							23			
22																							22			
21																							21			
20																							20			
19																							19			
18																							18			
17																							17			
16																							16			
15																							15			
14																							14			
13																							13			
12																							12			
11																							11			
10																							10			
9																							9			
8																							8			
7																							7			
6																							6			
5																							5			
4																							4			
3																							3			
2																							2			
1																							1			
0																							0			
fx																						N	$\Sigma =$			
x'	0	1	2	3	4	5	6	7	8	9	10	11	12	13	14	15	16	17	18	19	20					
fx'																						$\Sigma =$				
fx'^2																						$\Sigma =$				

X-Axis

Y-Axis	0	1	2	3	4	5	6	7	8	9	10	11	12	13	14	15	16	17	18	19	20	fy	y′	fy′	fy′²	x′y′
24																							24			
23																							23			
22																							22			
21																							21			
20																							20			
19																							19			
18																							18			
17																							17			
16																							16			
15																							15			
14																							14			
13																							13			
12																							12			
11																							11			
10																							10			
9																							9			
8																							8			
7																							7			
6																							6			
5																							5			
4																							4			
3																							3			
2																							2			
1																							1			
0																							0			
fx																						N	Σ=			
x′	0	1	2	3	4	5	6	7	8	9	10	11	12	13	14	15	16	17	18	19	20					
fx′																						Σ=				
fx′²																						Σ=				

Y-Axis	0	1	2	3	4	5	6	7	8	9	10	11	12	13	14	15	16	17	18	19	20	fy	y′	fy′	fy'^2	x′y′
24																							24			
23																							23			
22																							22			
21																							21			
20																							20			
19																							19			
18																							18			
17																							17			
16																							16			
15																							15			
14																							14			
13																							13			
12																							12			
11																							11			
10																							10			
9																							9			
8																							8			
7																							7			
6																							6			
5																							5			
4																							4			
3																							3			
2																							2			
1																							1			
0																							0			
fx																						N	Σ=			
x′	0	1	2	3	4	5	6	7	8	9	10	11	12	13	14	15	16	17	18	19	20					
fx′																						Σ=				
fx'^2																						Σ=				

X-Axis

Y-Axis	0	1	2	3	4	5	6	7	8	9	10	11	12	13	14	15	16	17	18	19	20	fy	y'	fy'	fy'²	x'y'
24																							24			
23																							23			
22																							22			
21																							21			
20																							20			
19																							19			
18																							18			
17																							17			
16																							16			
15																							15			
14																							14			
13																							13			
12																							12			
11																							11			
10																							10			
9																							9			
8																							8			
7																							7			
6																							6			
5																							5			
4																							4			
3																							3			
2																							2			
1																							1			
0																							0			
fx																						N	Σ=			
x'	0	1	2	3	4	5	6	7	8	9	10	11	12	13	14	15	16	17	18	19	20					
fx'																						Σ=				

Y-Axis	0	1	2	3	4	5	6	7	8	9	10	11	12	13	14	15	16	17	18	19	20	fy	y′	fy′	fy'^2	x′y′
24																							24			
23																							23			
22																							22			
21																							21			
20																							20			
19																							19			
18																							18			
17																							17			
16																							16			
15																							15			
14																							14			
13																							13			
12																							12			
11																							11			
10																							10			
9																							9			
8																							8			
7																							7			
6																							6			
5																							5			
4																							4			
3																							3			
2																							2			
1																							1			
0																							0			
fx																						N	Σ=			
x′	0	1	2	3	4	5	6	7	8	9	10	11	12	13	14	15	16	17	18	19	20					
fx′																						Σ=				
fx'^2																						Σ=				

X-Axis

Y-Axis	0	1	2	3	4	5	6	7	8	9	10	11	12	13	14	15	16	17	18	19	20	fy	y′	fy′	fy'^2	x′y′
24																							24			
23																							23			
22																							22			
21																							21			
20																							20			
19																							19			
18																							18			
17																							17			
16																							16			
15																							15			
14																							14			
13																							13			
12																							12			
11																							11			
10																							10			
9																							9			
8																							8			
7																							7			
6																							6			
5																							5			
4																							4			
3																							3			
2																							2			
1																							1			
0																							0			
fx																						N	Σ=			
x′	0	1	2	3	4	5	6	7	8	9	10	11	12	13	14	15	16	17	18	19	20					
fx′																						Σ=				
fx'^2																						Σ=				

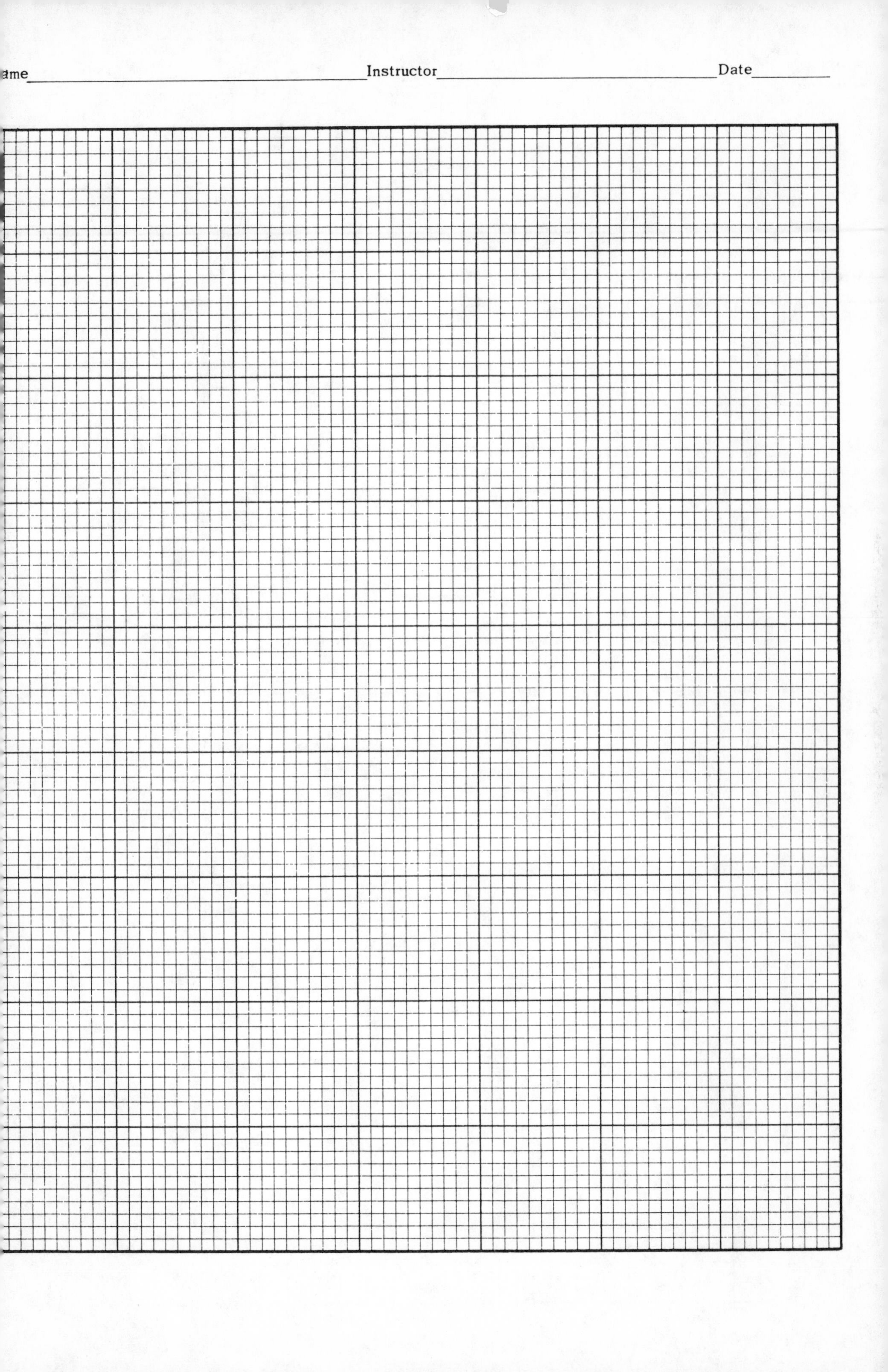
ame
Instructor
Date

Name__________ Instructor__________ Date__________

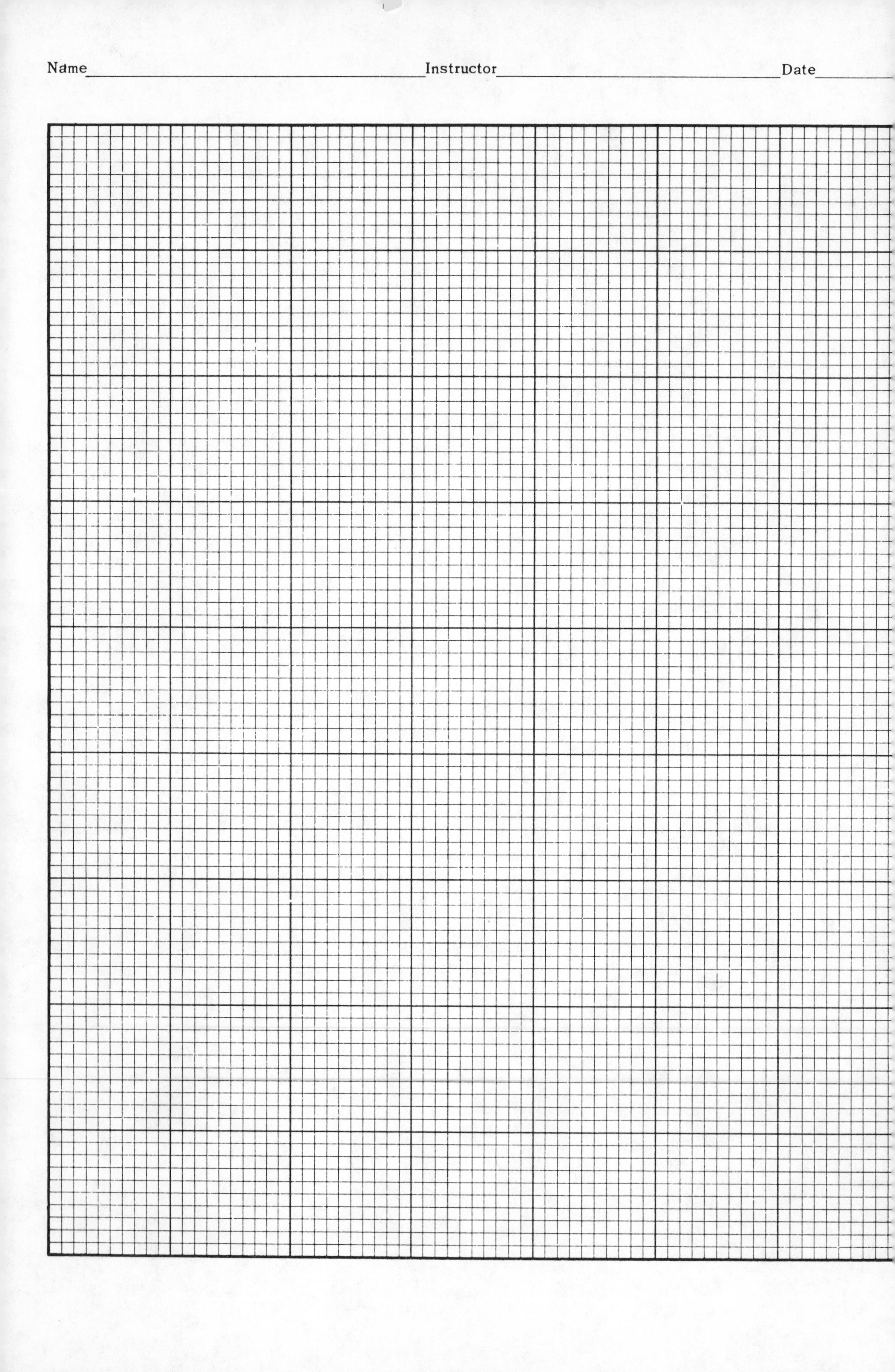

ame______ Instructor______ Date______

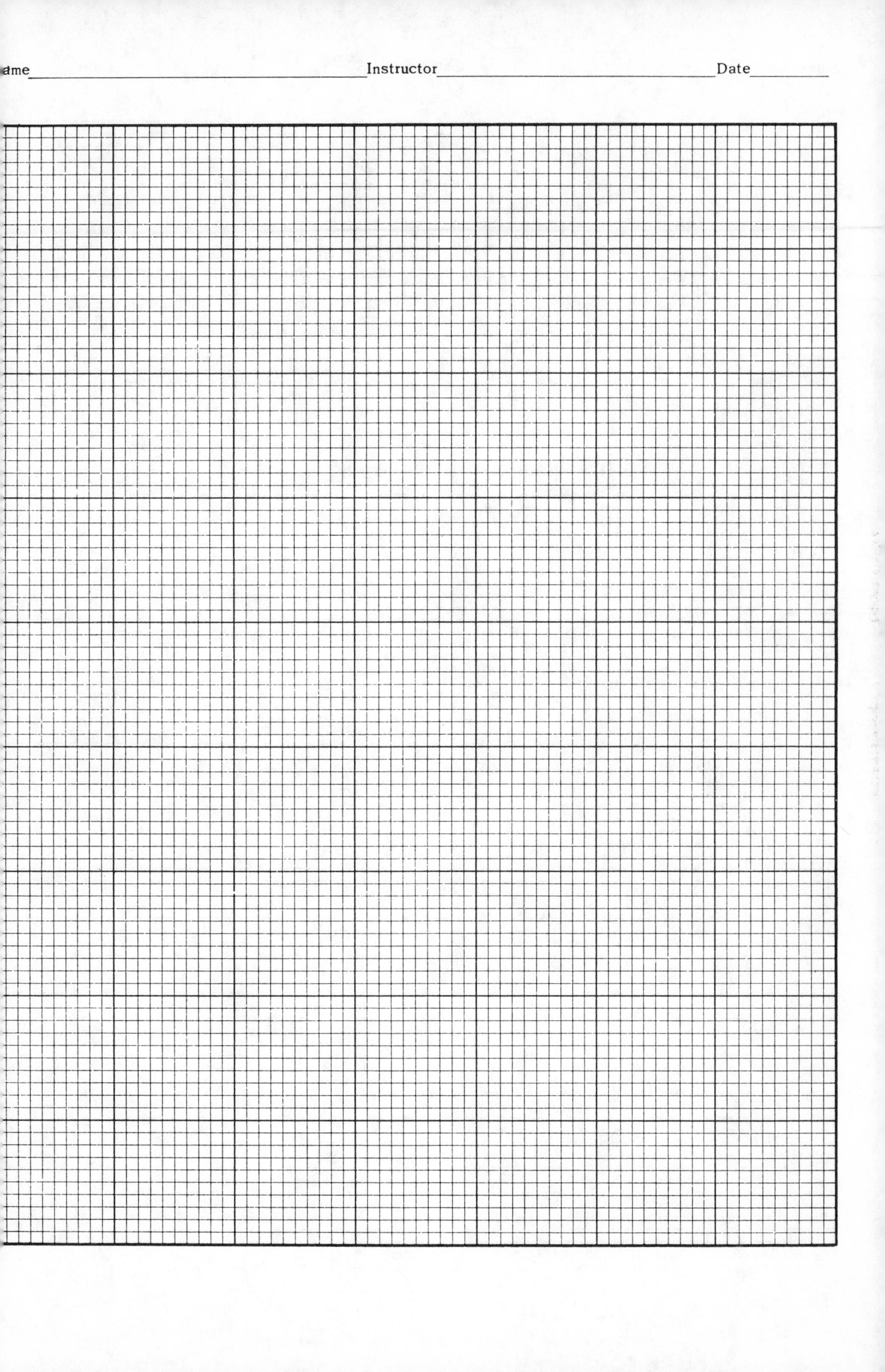

Name______________________Instructor______________________Date__________

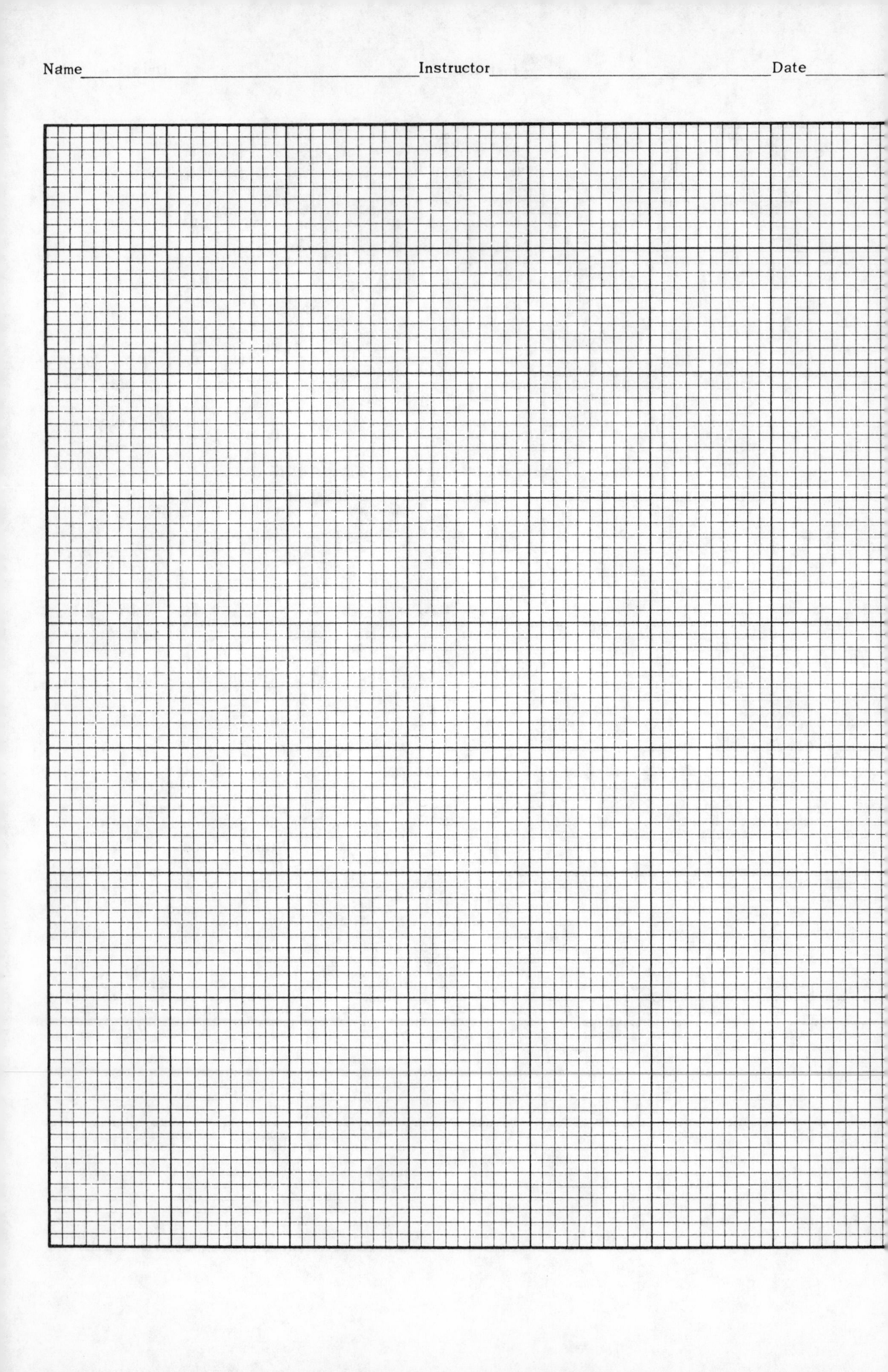

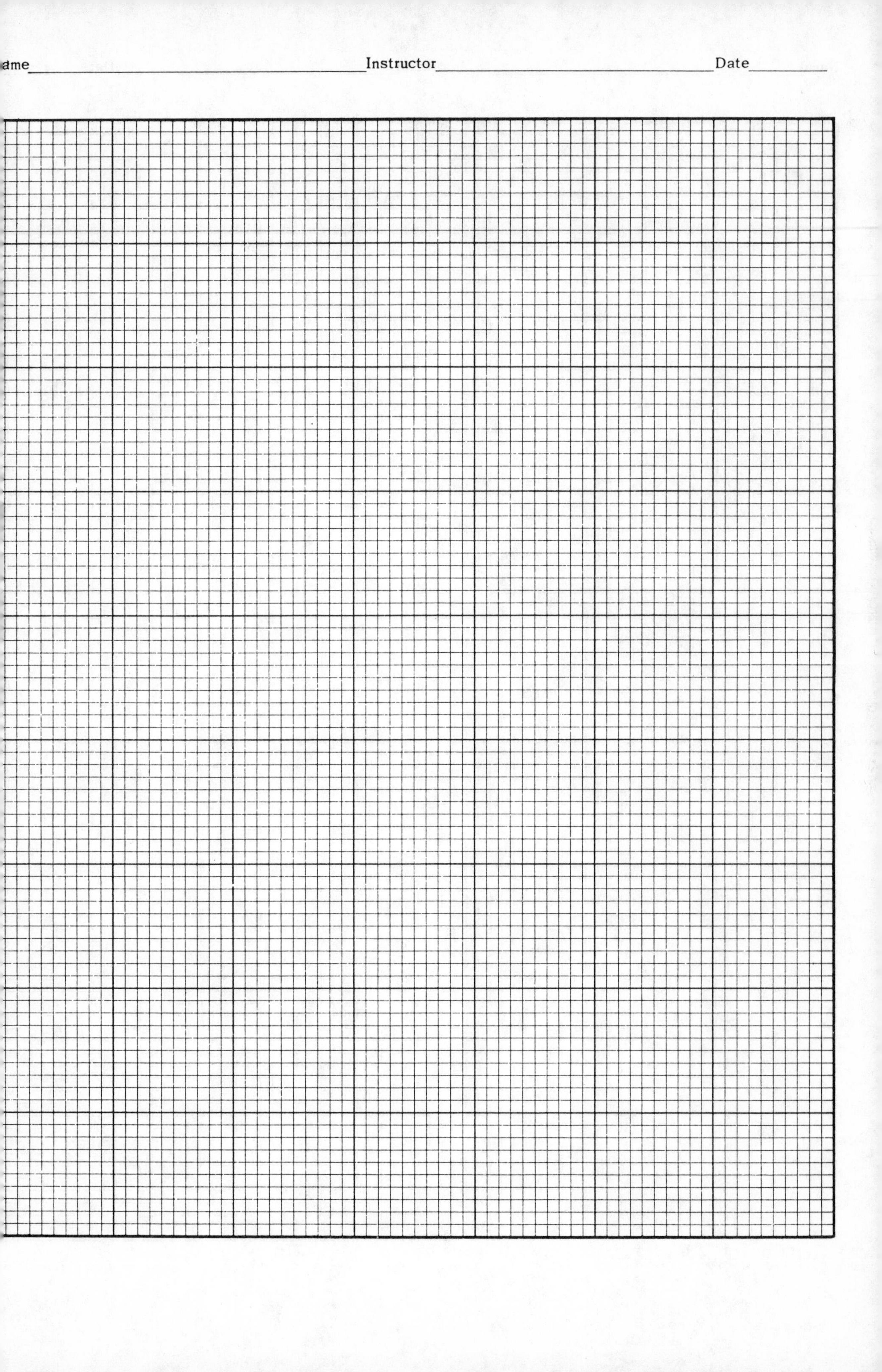
ame
Instructor
Date

Name__________ Instructor__________ Date__________

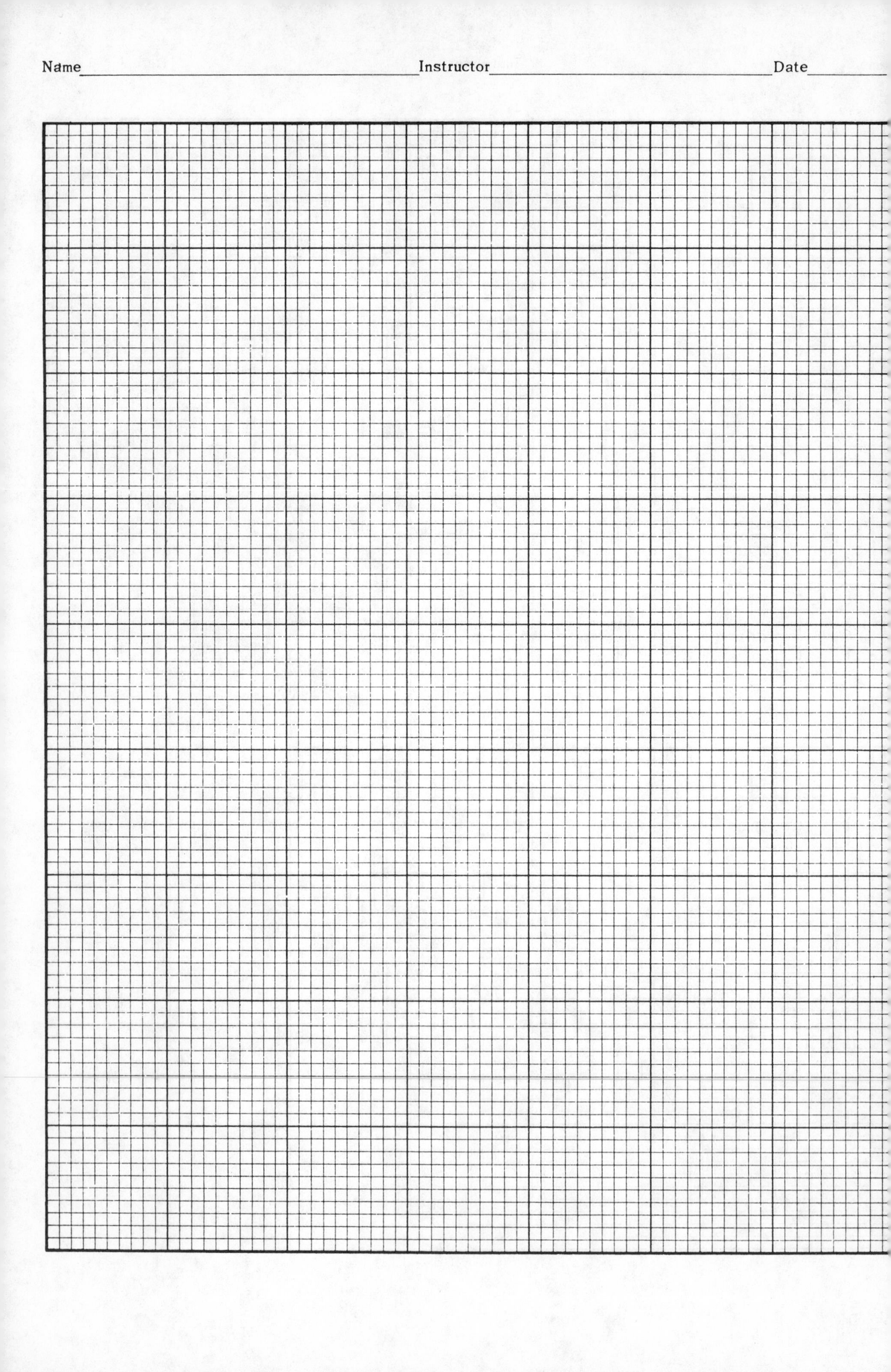

me______________________________ Instructor______________________________ Date__________

Name______________________ Instructor______________________ Date__________

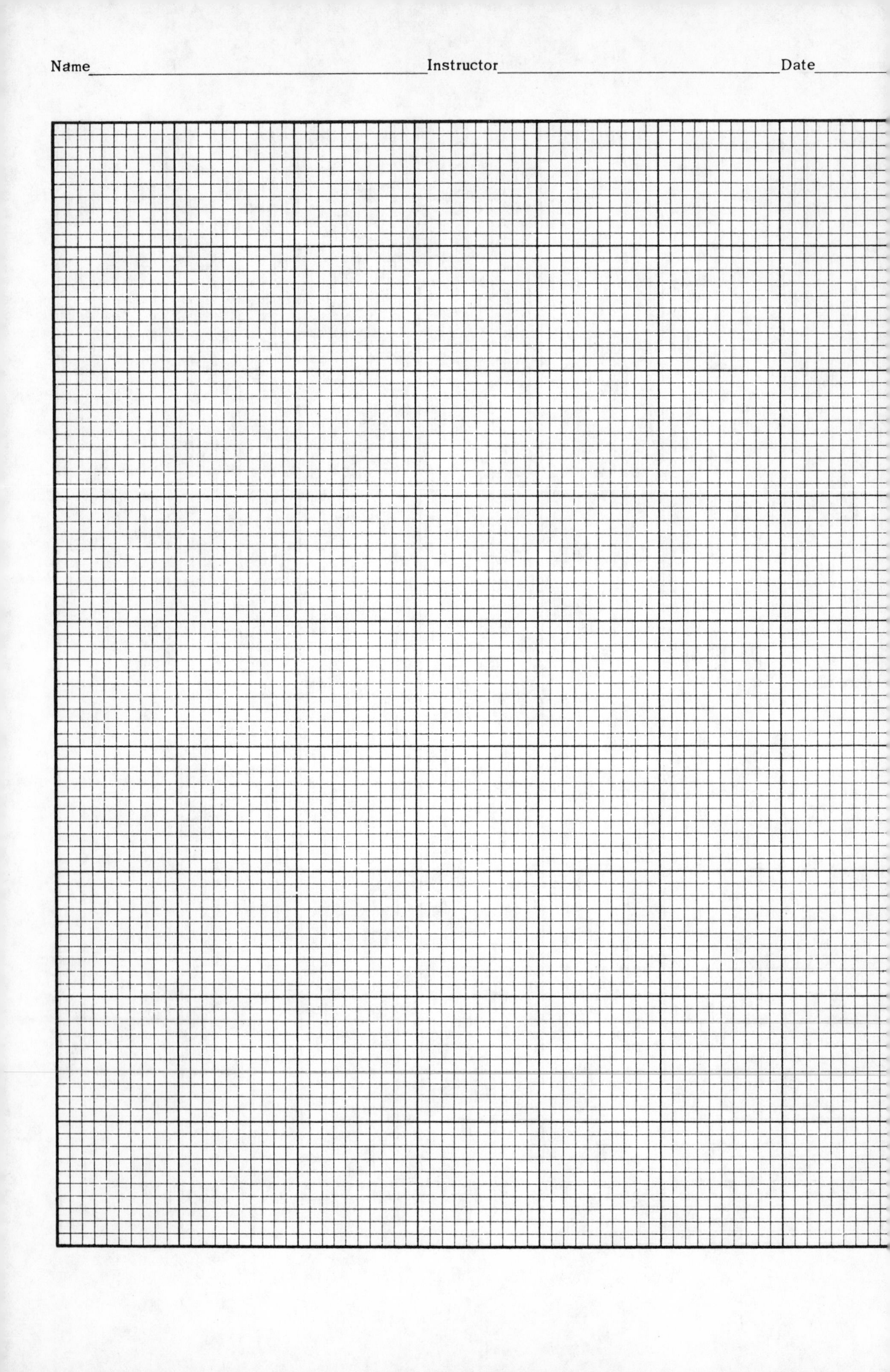

me______________________ Instructor______________________ Date__________

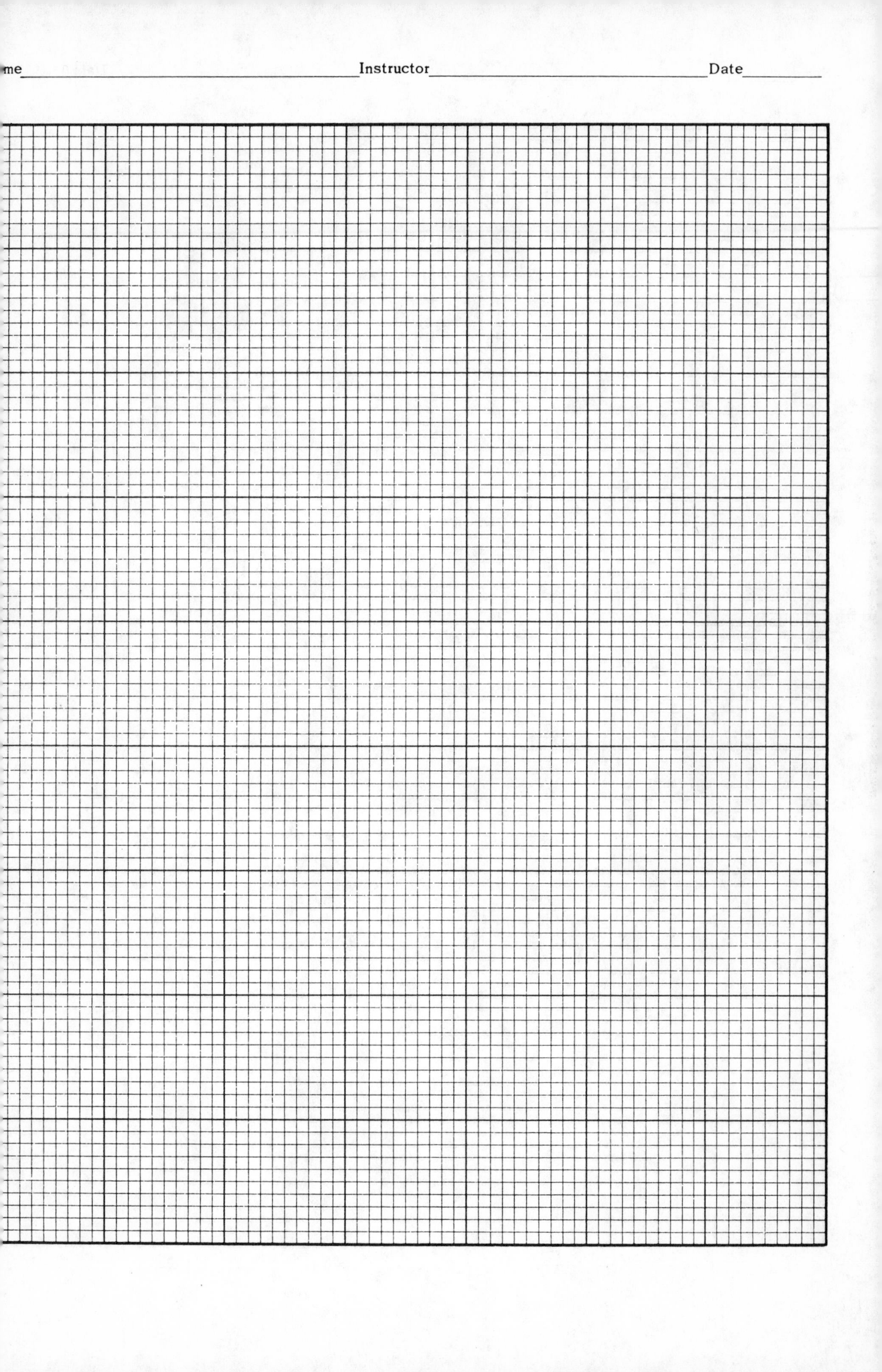

Name______________________ Instructor______________________ Date__________

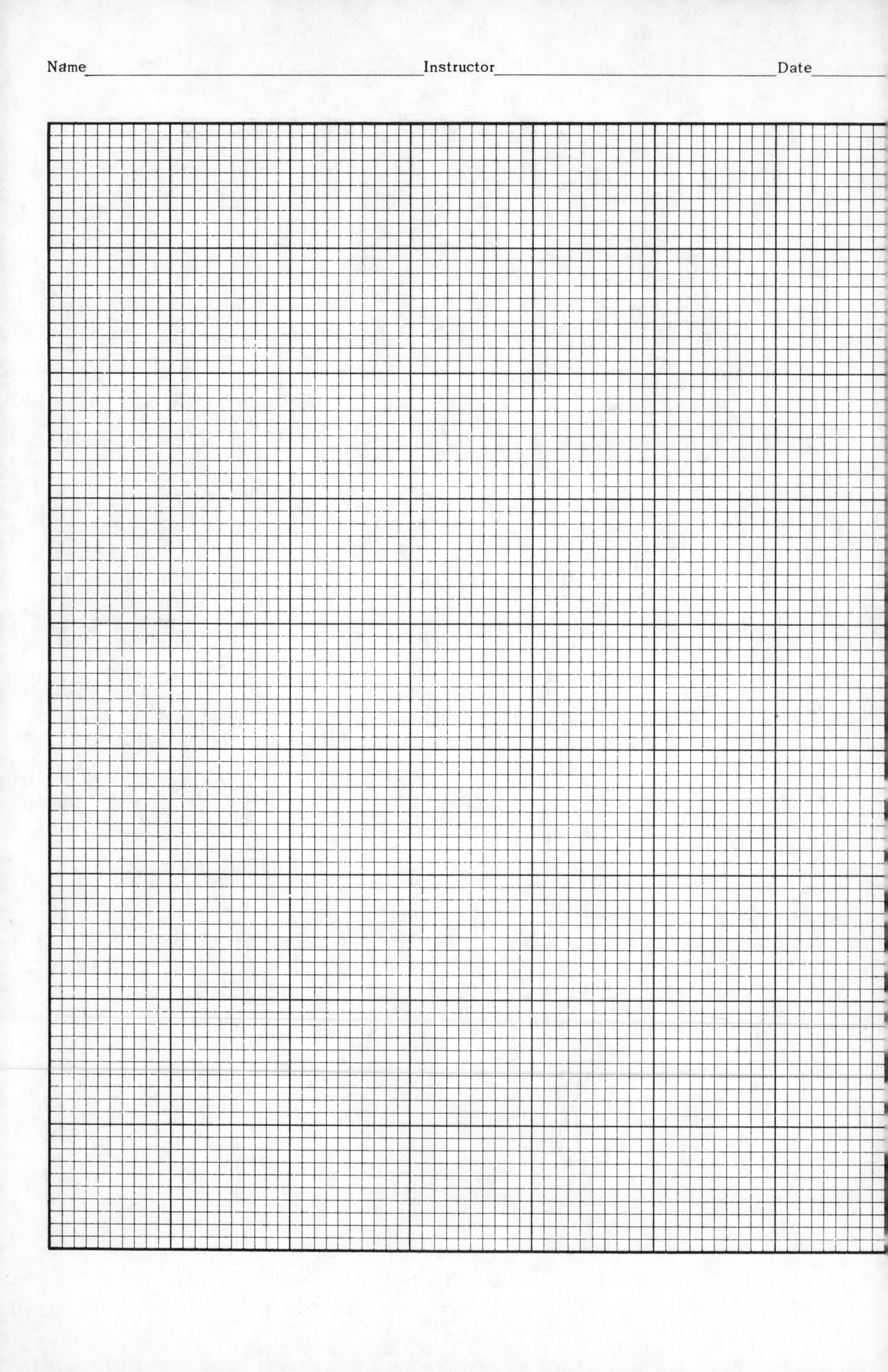

me______________ Instructor______________ Date__________

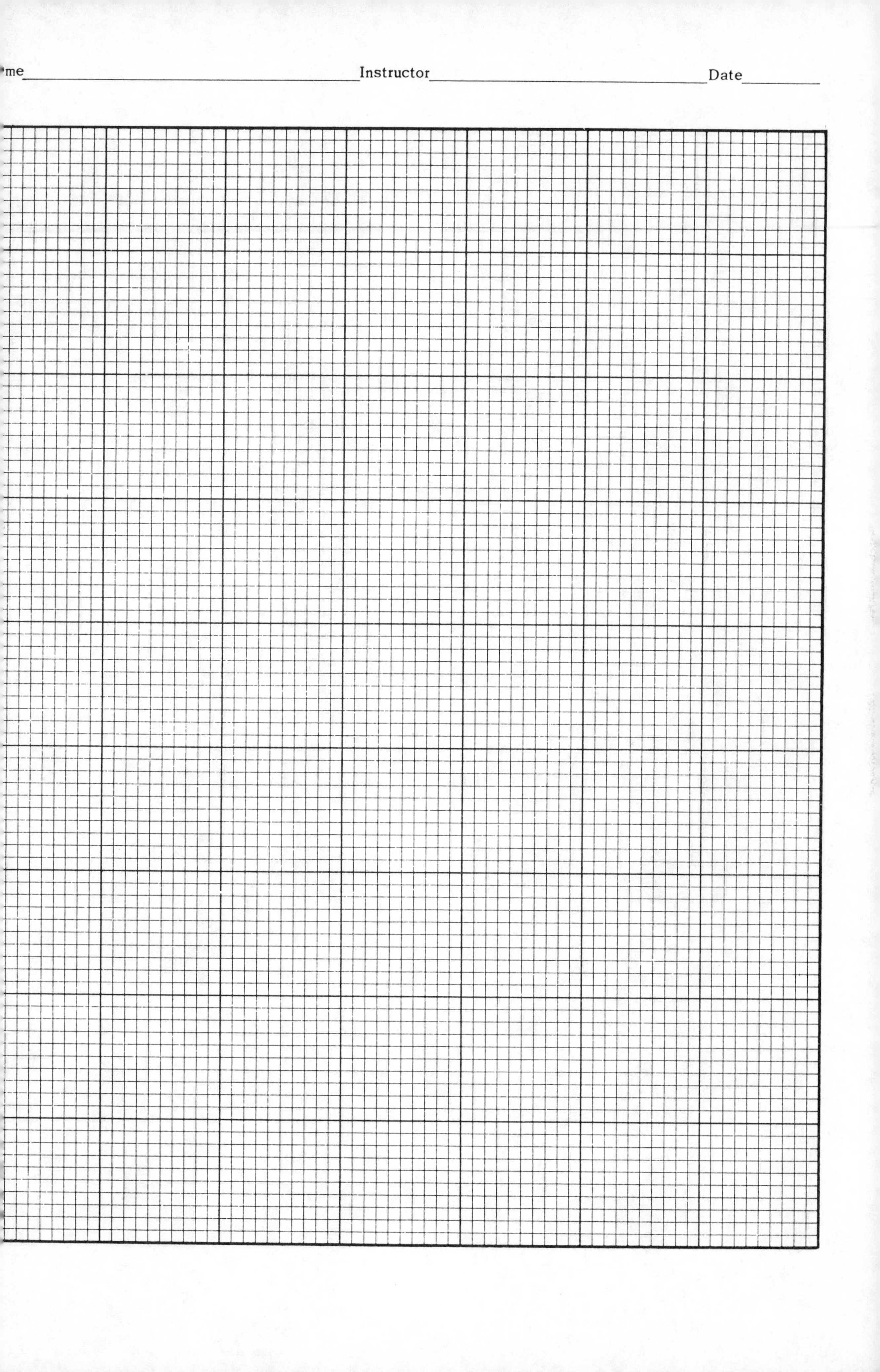

Name______________________________ Instructor______________________________ Date__________

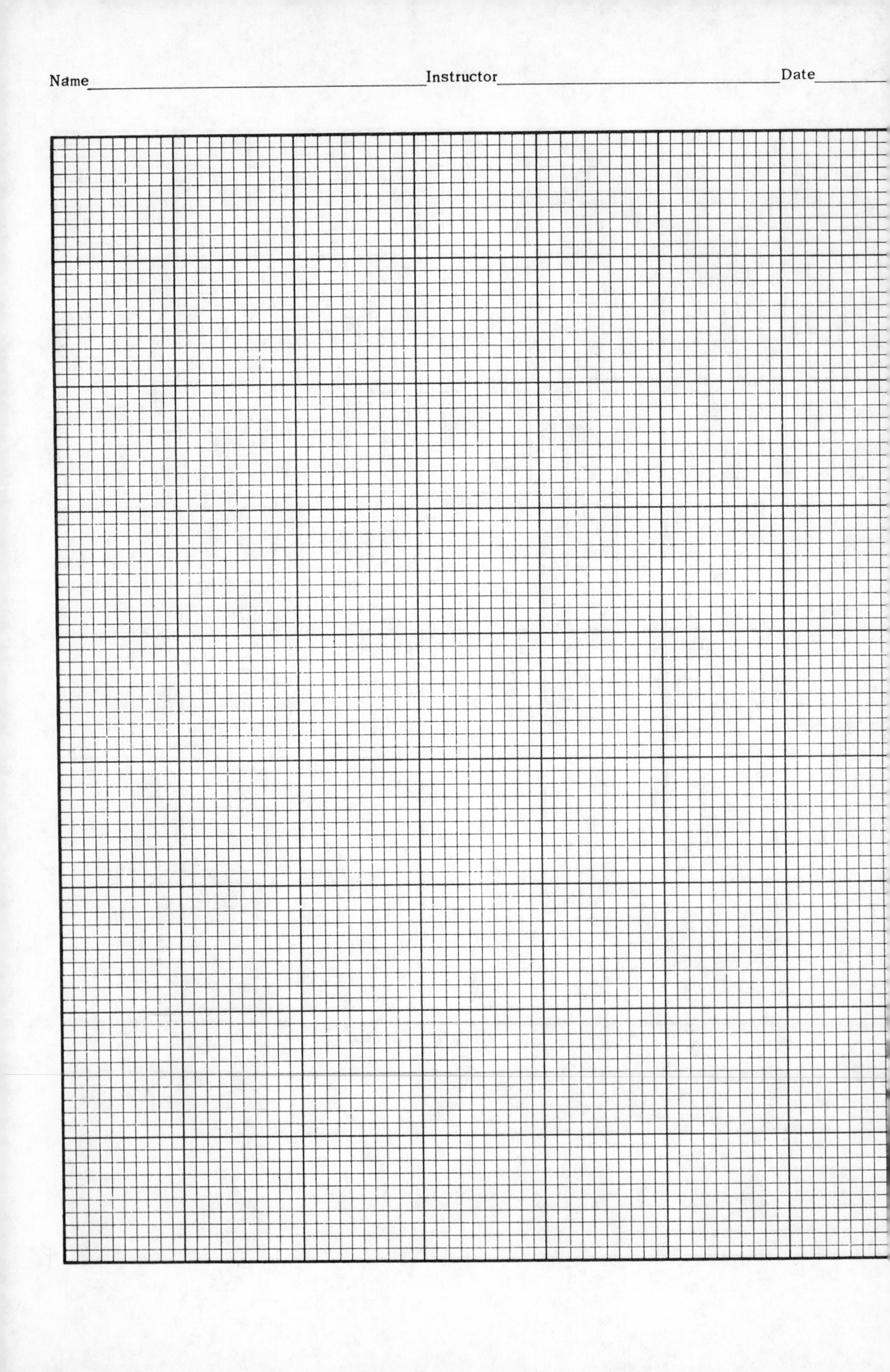

me______________________________ Instructor______________________________ Date__________

Name______________________________Instructor______________________________Date__________

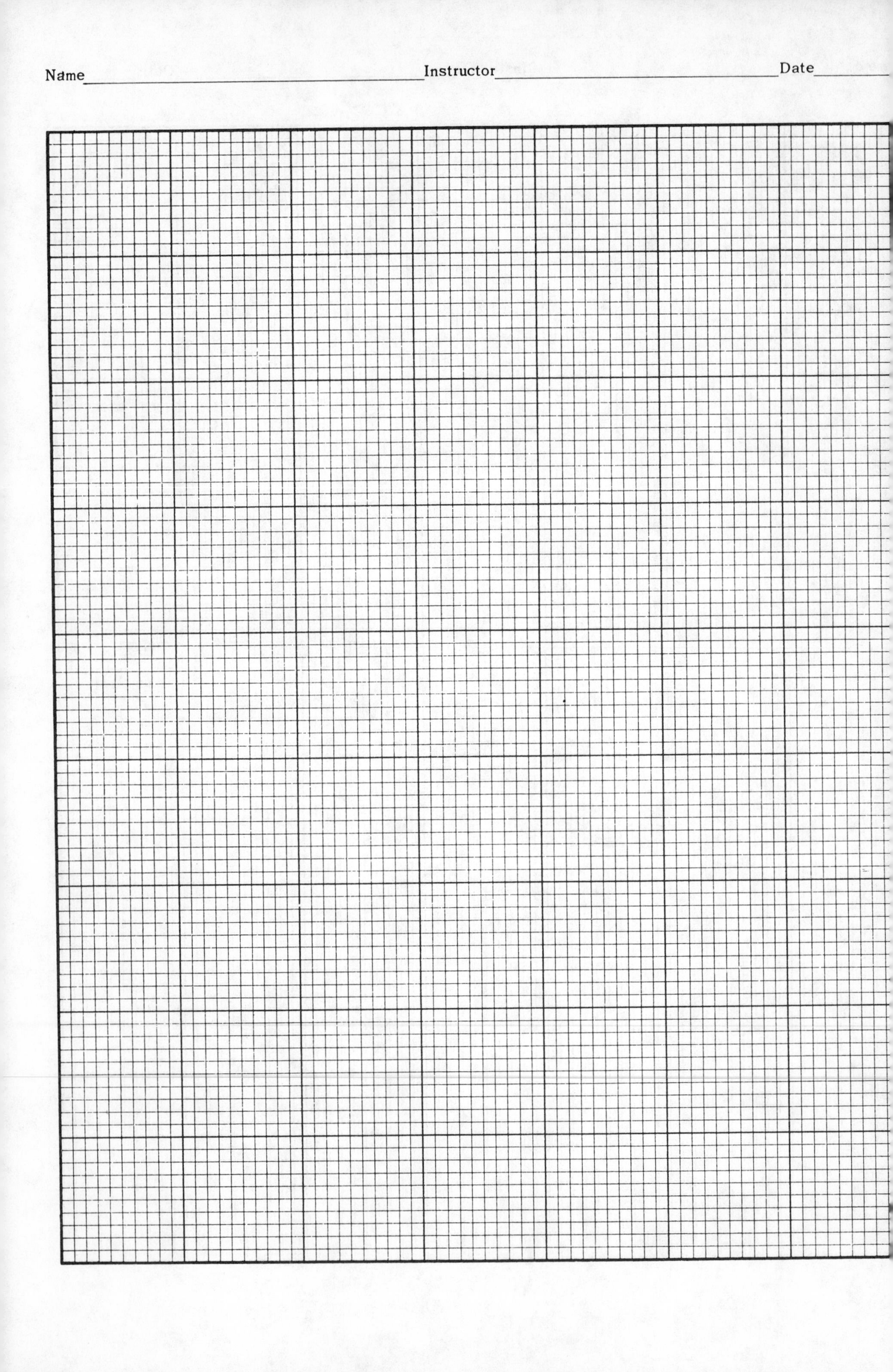

ame________________________________ Instructor________________________________ Date__________

Name______________________Instructor______________________Date________

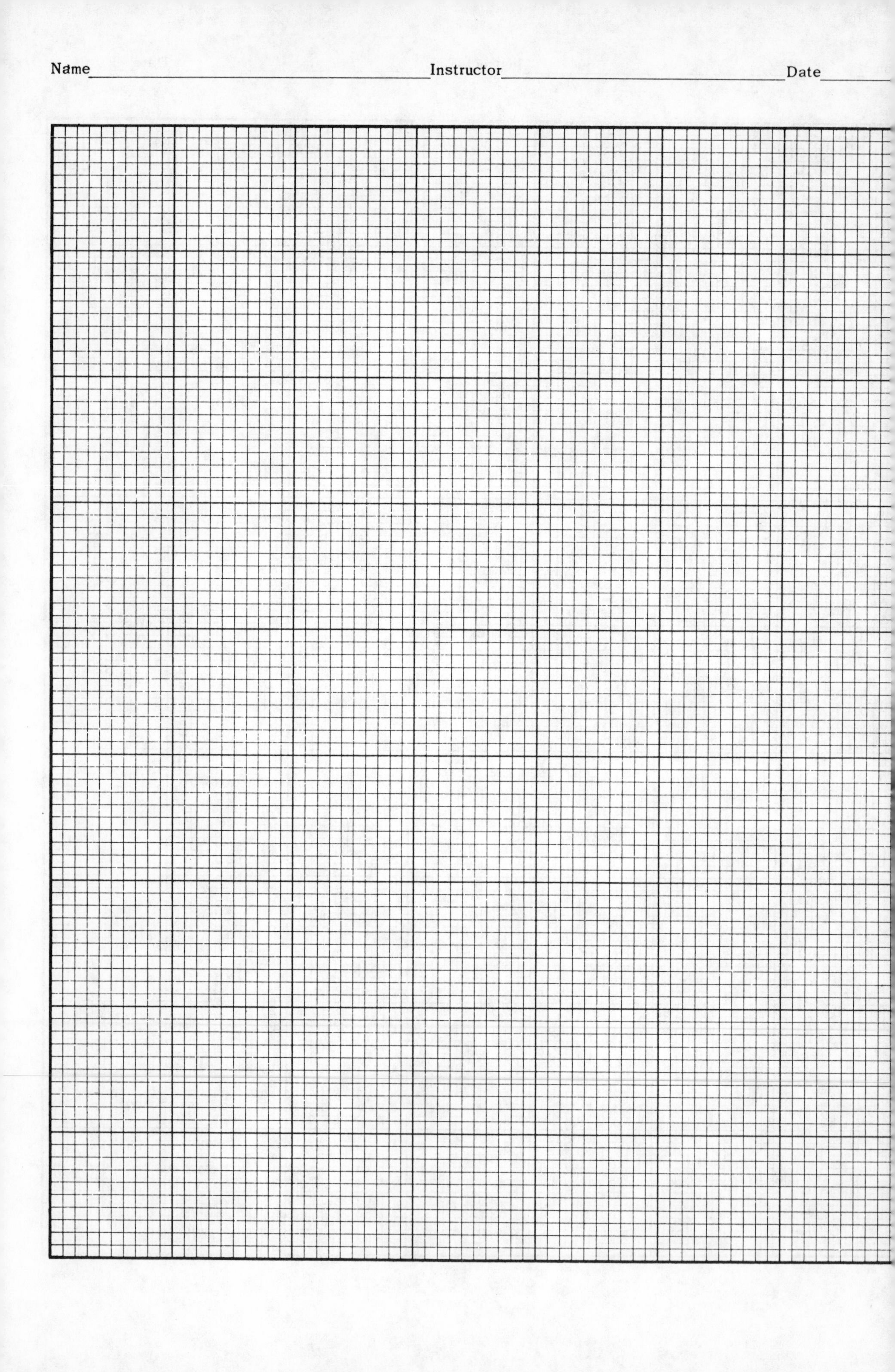

ame______ Instructor______ Date______

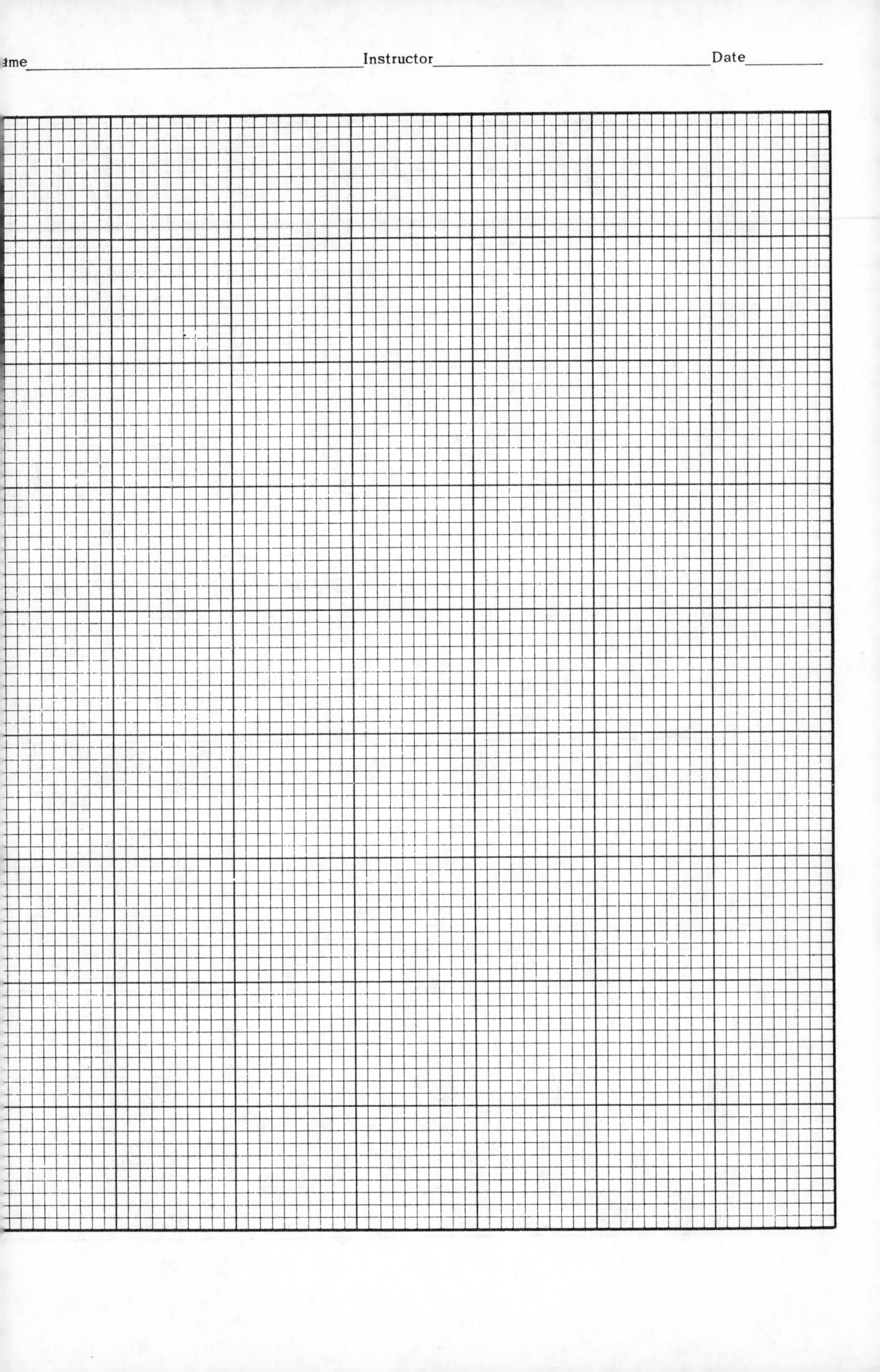

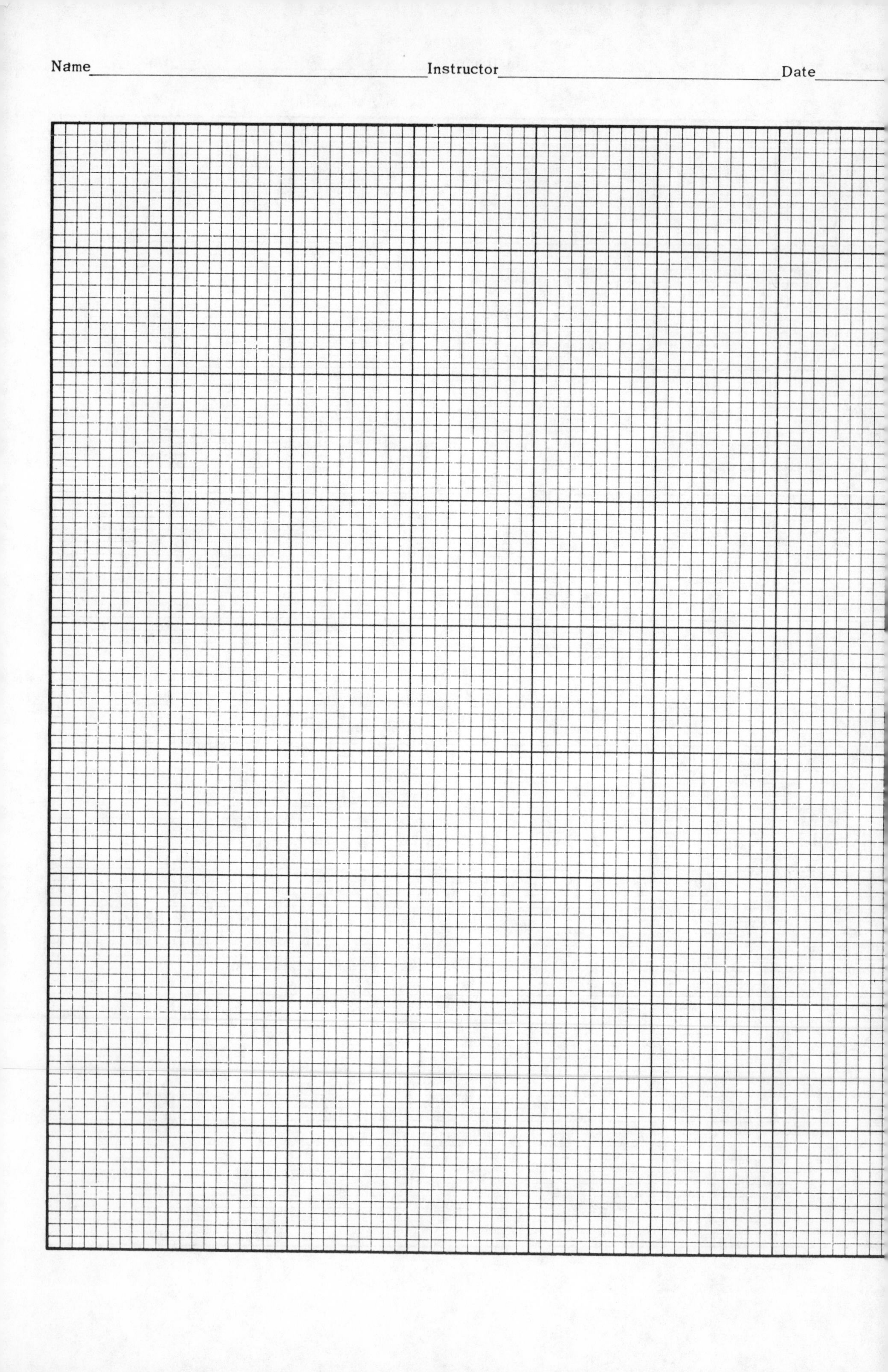
Name
Instructor
Date

me______________________ Instructor______________________ Date__________

Name________________________________ Instructor________________________________ Date__________

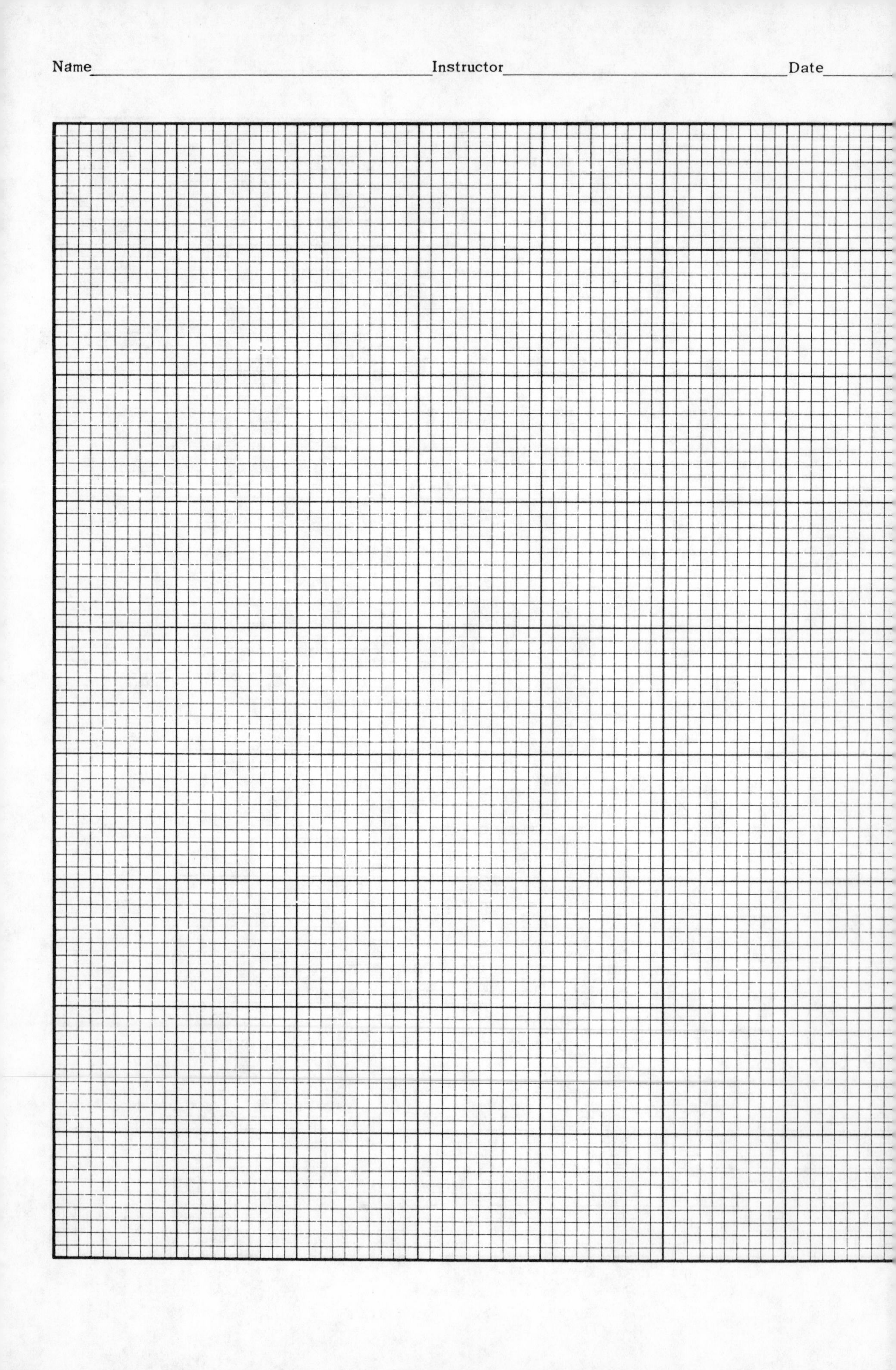

me____________________ Instructor____________________ Date________

75 76 9 8 7 6 5 4 3